Lecture Notes in Physics

T0238571

Springer

Berlin
Heidelberg
New York
Barcelona
Hong Kong
London
Milan
Paris
Singapore
Tokyo

Physics and Astronomy

ONLINE LIBRARY

http://www.springer.de/phys/

The Editorial Policy for Monographs

The series Lecture Notes in Physics reports new developments in physical research and teaching - quickly, informally, and at a high level. The type of material considered for publication includes monographs presenting original research or new angles in a classical field. The timeliness of a manuscript is more important than its form, which may be preliminary or tentative. Manuscripts should be reasonably self-contained. They will often present not only results of the author(s) but also related work by other people and will provide sufficient motivation, examples, and applications.

Acceptance

The manuscripts or a detailed description thereof should be submitted either to one of the series editors or to the managing editor. The proposal is then carefully refereed. A final decision concerning publication can often only be made on the basis of the complete manuscript, but otherwise the editors will try to make a preliminary decision as definite as they can on the basis of the available information.

Contractual Aspects

Authors receive jointly 30 complimentary copies of their book. No royalty is paid on Lecture Notes in Physics volumes. But authors are entitled to purchase directly from Springer-Verlag other books from Springer-Verlag (excluding Hager and Landolt-Börnstein) at a $33\frac{1}{3}$% discount off the list price. Resale of such copies or of free copies is not permitted. Commitment to publish is made by a letter of interest rather than by signing a formal contract. Springer-Verlag secures the copyright for each volume.

Manuscript Submission

Manuscripts should be no less than 100 and preferably no more than 400 pages in length. Final manuscripts should be in English. They should include a table of contents and an informative introduction accessible also to readers not particularly familiar with the topic treated. Authors are free to use the material in other publications. However, if extensive use is made elsewhere, the publisher should be informed. As a special service, we offer free of charge LaTeX macro packages to format the text according to Springer-Verlag's quality requirements. We strongly recommend authors to make use of this offer, as the result will be a book of considerably improved technical quality. The books are hardbound, and quality paper appropriate to the needs of the author(s) is used. Publication time is about ten weeks. More than twenty years of experience guarantee authors the best possible service.

LNP Homepage (http://www.springerlink.com/series/lnp/)

On the LNP homepage you will find:
−The LNP online archive. It contains the full texts (PDF) of all volumes published since 2000. Abstracts, table of contents and prefaces are accessible free of charge to everyone. Information about the availability of printed volumes can be obtained.
−The subscription information. The online archive is free of charge to all subscribers of the printed volumes.
−The editorial contacts, with respect to both scientific and technical matters.
−The author's / editor's instructions.

Mirko Degli Esposti   Sandro Graffi   (Eds.)

The Mathematical Aspects of Quantum Maps

 Springer

Editors

Mirko Degli Esposti
Sandro Graffi
Università di Bologna
Dipartimento di Matematica
Piazza di Porta San Donato 5
40127 Bologna, Italy

Cataloging-in-Publication Data applied for

A catalog record for this book is available from the Library of Congress.

Bibliographic information published by Die Deutsche Bibliothek

Die Deutsche Bibliothek lists this publication in the Deutsche Nationalbibliografie;
detailed bibliographic data is available in the Internet at http://dnb.ddb.de

ISSN 0075-8450
ISBN 978-3-642-05699-4   e-ISBN 978-3-540-37045-1

Springer-Verlag Berlin Heidelberg New York
a member of BertelsmannSpringer Science+Business Media GmbH

http://www.springer.de

© Springer-Verlag Berlin Heidelberg 2010
Printed in Germany

Camera-data conversion by Steingraeber Satztechnik GmbH Heidelberg
Cover design: *design & production*, Heidelberg

Printed on acid-free paper
55/3141/du - 5 4 3 2 1 0

Foreword

Quantum chaos usually means the investigation of quantum mechanical properties of classically chaotic systems. Expressed differently: what are the manifestations, if any, of chaotic behavior in classical systems on their quantum counterparts?

This question has obviously been thoroughly investigated since the early days of quantum mechanics, for instance, in the classical papers of Von Neumann on quantum ergodicity [14] and of Wigner on the quantum analog of the Boltzmann equation [15]. More recent approaches to this general question can be found for instance in [8], [9] and [10].

This kind of issue has obvious foundational aspects. For example, the need to justify quantum statistical mechanics in the same way as classical chaos is the basis for justifying classical statistical mechanics. However, investigations in quantum chaos also arise from questions of direct and important physical relevance: the spectral statistics in complex nuclei and the "stochastic" ionization of the hydrogen atom in a microwave field, to mention only two. The study of spectral statistics led to the formulation of standard conjectures on the probability distribution of the eigenvalue spacing. The Poisson probability distribution should be valid for the quantized counterpart of the classically integrable systems (Berry-Tabor conjecture, formulated in the seventies [4], see also [11,12]) and the COE/CUE (Circular Orthogonal/Unitary Ensemble) probability distribution [30] for the quantized counterpart of the classically chaotic systems (Bohigas-Giannoni-Schmit conjecture, formulated in the eighties [1]). The numerical evidence for the validity of this conjecture is so convincing that the spacing distribution among quantum levels is considered as a characteristic property of quantum chaotic systems even though no example is known where a proof can be obtained (see for example [8,10] and references therein). Usually the spacing distribution is explored through some spectral indicators such as the form factor (namely the Fourier transform of the two-point correlation function), according to the COE/CUE statistics (yielding the so-called level repulsion)

On the other hand, the study of time-dependent models (such as the kicked rotator) or other simplified versions of the hydrogen atom in a microwave field produced a second important numerical discovery, the quantum suppression of classical chaos. The numerical evidence (mainly on the kicked rotator and its variants) shows that classical diffusive behavior at large times is transformed into localized behavior after quantization ([6,7], for a general review see [5] and references therein).

The mathematical understanding of these and many other, "quantum chaotic" features is not easy for several conceptual and technical reasons. First, it has to be recalled that chaotic behavior in classical systems is described in terms of well-defined mathematical notions derived from ergodic theory (ergodicity, mixing, K-property, exponential decay of correlations, etc.), which ultimately depend on the possibility of localizing the initial conditions with arbitrary accuracy. Typically, the sensitive dependence on initial conditions is the standard dynamical mechanism generating chaotic behavior. The indeterminacy principle altogether excludes this mechanism in quantum mechanics and, moreover, makes the classical limit a much subtle problem in quantized "chaotic" systems than in quantized integrable ones. Another way of looking at the nontriviality of the classical limit in this situation is the noncommutativity of the two limits $t \to \infty$ and $\hbar \to 0$ because the very definition of chaotic behavior in classical systems actually requires the limit $t \to \infty$.

The iteration of area preserving maps, which also sometimes describe Hamiltonian systems at discrete times, represent the basic examples of ergodic theory (the Arnold cat, the baker's transformation and the sawtooth map, just to mention few) and therefore the most thoroughly studied examples of non-trivial discrete-time evolution which may give rise to chaotic behavior. Since the phase space of classical discrete dynamics is compact, again by the indeterminacy principle its quantum counterpart will allow only a finite number of states (roughly speaking, a quantum state occupies a volume of size at least $\hbar$) and therefore the quantum evolution takes place within a finite dimensional Hilbert space. The classical limit corresponds to the Hilbert space dimension tending to infinity. The investigation of the quantum maps requires, therefore, the preliminary clarification of several nontrivial mathematical questions, beginning from an unambiguous implementation of a quantization procedure. It will be described how the quantization defines the quantum evolution as the iteration of a unitary $N \times N$ matrix (the propagator), N being proportional to the inverse Planck constant. The quantized maps thus represent the simplest testing ground for the verification not only of the above mentioned conjecture on spectral statistics, but also for novel concepts isolated in the context of the classical limit of quantized chaotic systems, such as (unique) quantum ergodicity, and "scarring" of the eigenfunctions. Moreover, quantum maps is one of the most relevant physical examples where techniques from analytic and algebraic number theory emerge naturally.

Finally, it is also important to notice that quantum maps, seen as unitary transformations on finite Hilbert spaces (of dimension $N = 2^n$) could be interpreted as the implementation of particular quantum gate acting on the space of qubits, as defined in the theory of quantum computation. The exploration of the mathematical properties of quantum maps could shed some light on the understanding of the implementation of new and more efficient quantum algorithms.

The present volume deals precisely with the mathematical aspects of quantum maps mentioned above. Its purpose is to give a broad and in-depth overview of the mathematical problems arising in quantum maps, with a particular emphasis on their physical origin and on their numerical investigation, from the beginning (which may be traced back to the classical 1980 paper of Berry and Hannay on the quantization of linear hyperbolic symplectic maps over the torus [3] and to the 1983 paper by Balazs and Voros on the quantization of the baker map [4]) to the present state of the art. The contents of the present volume reflects the approach chosen to achieve that purpose. Three main topics are covered in full detail. The first topic presented , by A. Knauf, is an overview of that part of classical dynamical systems most relevant to quantum chaos in general and quantum maps in particular, mainly ergodic theory. The second topic, by Z.Rudnick, is an introductory presentation of some basic concepts and techniques out of number theory.

The third topic, by the editors of this book, concerns the general quantization procedure, including observables, for the symplectic maps, both linear (e.g. the Arnold cat) as well as piecewise linear (e.g. the baker's transformation) and the related mathematical results concerning the recovery of the classical chaotic behavior at the classical limit. A fourth contribution, by A. Bäcker, is an introduction to the the numerical aspects in quantum chaos, in particular eigenvalue and eigenfunction computations for chaotic quantum systems, such as discrete maps or two-dimensional billiards. The classical dynamics of two-dimensional area-preserving maps on the torus is illustrated using the standard map and a perturbed cat map. The quantization of area-preserving maps given by their generating function is discussed and for the computation of the eigenvalues a computer program in Python is presented. In particular, the author illustrates the eigenvalue distribution for two types of perturbed cat maps, one leading to COE and the other to CUE statistics. For the eigenfunctions of quantum maps, the author studies the distribution of the eigenvectors and compare them with the corresponding random matrix distributions.

The final contribution, by R. Artuso, is meant to illustrate some features of deterministic transport in chaotic systems. The first part of this contribution deals with the case of *hyperbolic* systems, where typically normal diffusion is observed, while the second part explores *weakly* chaotic systems, where long trappings near regular phase-space regions may induce anomalies in diffusive properties. Examples of analytic calculations are given in the framework of *cycle expansions*, a general technique for obtaining chaotic averages.

References

1. O. Bohigas, M.-J. Giannoni and C. Schmit, *Characterization of chaotic quantum spectra and universality of level fluctuation laws* Phys. Rev. Lett. **52** , 1–4 (1984).

2. M.V. Berry and J.H. Hannay, *Quantization of linear maps on a torus – Fresnel diffraction by a periodic grating*, Physica D **1** (1980), 267–291.

3. N.L. Balazs, A. Voros, *The quantized Baker's transformation*, Annals of Physics **190** (1989), 1–31.

4. M.V. Berry and M. Tabor *Level clustering in the regular spectrum* Proc. Roy. Soc. A **356**, 375-394 (1977).

5. G. Casati and J. Ford *Stochastic behavior in classical and quantum Hamiltonian systems*, Proc. Como Conf. 1977. Lect. Notes Physics **93**, Springer-Verlag Berlin Heidelberg New York (1977).

6. B.V. Chirikov and G.M. Zaslavskii, Sov. Phys. Uspekhi **14**, 549 (1972).

7. B.V. Chirikov and G.M. Zaslavskii, Sov. Phys. JETP **46**, 1094 (1977).

8. M.C. Gutzwiller *Chaos in classical and quantumMechanics*. Springer-Verlag Berlin Heidelberg New York, IAM 1, (1990).

9. M.-J. Giannoni, A. Voros and J. Zinn-Justin *Chaos and Quantum Physics: Les Houches 1989*, North-Holland (1991).

10. F. Haake *Quantum signatures of chaos*, Springer-Verlag Berlin Heidelberg New York (2001).

11. J. Marklof *Level spacing statistics and integrable dynamics*, Proceedings of the XIIIth International Congress on Mathematical Physics, London 2000, 359–363 (2001).

12. J. Marklof *The Berry-Tabor conjecture*, Proceedings of the 3rd European Congress of Mathematics, Barcelona 2000, Progress in Mathematics **202**, 421–427 (2001).

13. M.L. Mehta *Random matrices*. Academic Press,New York and London, (1991).

14. J.von Neumann *Beweis des Ergodensatzes und des H-Theorems in der Neuen Mechanik* Zschr.f.Physik **57**, 30–70 (1929).

15. E.P. Wigner, Phys. Rev. **40**, 749 (1932).

Preface

The present volume represents the collected lectures of a Summer School on mathematical aspects of quantum maps held at Universitá di Bologna (Villa Gandolfi-Pallavicini) from September 1 to September 11, 2001.

The Summer School has been organized as an institutional activity of the EC-supported European Research and Training Network "The Mathematics of Quantum Chaos", whose nodes are at the Universities of Bristol, UK (co-ordinating node), Bologna (Italy), Paris XI (France), Tel Aviv (Israel), Ulm (Germany) and Uppsala (Sweden). Its purpose was to give a broad and in-depth overview of the mathematical problems arising in quantum maps, with a particular emphasis on their physical origin and numerical investigation.

Acknowledgement

This work has been supported by the European Commission under the Research Training Network (Mathematical Aspects of Quantum Chaos) no HPRN-CT-2000-00103 of the IHP Programme.

Bologna, Italy
January 2003

Mirko Degli Esposti
Sandro Graffi

Contents

Introduction to Dynamical Systems

Andreas Knauf[1]

Mathematisches Institut, Universität Erlangen-Nürnberg, Bismarckstr. $1\frac{1}{2}$, D-91054 Erlangen, Germany, *knauf@mi.uni-erlangen.de*

1 Introduction

Isaac Newton had one insight that he considered to be so fundamental that he kept it secret:

"It is useful to solve differential equations."[1]

The notion of dynamical systems grew out of the theory of differential equations. It was realized by Henri Poincaré that equations like the one of the celestial three body problem could not be solved analytically. Thus it was necessary to supplement the quantitative approximate solutions by qualitative methods in order to understand the long time behaviour of the solutions of differential equations.

Given an ordinary differential equation (O.D.E.) in explicit form, one may reduce it to a time-independent O.D.E. of first order

$$\dot{x}(t) \equiv \frac{d}{dt}x(t) = V(x(t)),$$

where x is a point in *phase space* M. In the simplest case M is an open subset of $\mathbb{R}^d$, and the *vector field* $V : M \to \mathbb{R}^d$ is a smooth map.

$t \in \mathbb{R}$ is the *time* parameter, and ideally we want to have a solution $t \mapsto x(t; x_0)$ of the *initial value problem*

$$\dot{x} = V(x) \quad , \quad x(0) = x_0$$

for all *initial values* $x_0 \in M$ and times $t \in \mathbb{R}$. In that case we obtain, geometrically speaking, a flow

$$\Phi_t : M \to M \qquad (t \in \mathbb{R})$$

$\Phi_t(x_0) = x(t, x_0)$ which is a smooth one-parameter family of maps of phase space, meeting the equations

$$\Phi_0 = \mathrm{Id}_M \quad , \quad \Phi_{t_1} \circ \Phi_{t_2} = \Phi_{t_1+t_2}.$$

There are now two ways to study the dynamical system Φ_t:

---

[1] Cited from V. Arnold: Geometrical Methods in the Theory of Ordinary Differential Equations. Springer 1983

- One may explore its *topological properties*.
  As an example, one may ask whether $\lim_{t\to\infty} \Phi_t(x_0)$ exists for a given initial condition $x_0 \in M$.
- Alternatively, one may study properties of Φ_t that are typical from a measure theoretical view, i.e. its *ergodic properties*.

We shall now concentrate on the second question. Lateron, we shall study geometric properties of flows giving rise to ergodicity. The recent volume [Bu] of the *Encyclopaedia of Mathematical Sciences* provides an overview.

2 Measure-Preserving Dynamical Systems

Ergodic theory studies measure preserving group actions. In the O.D.E. setting discussed above the group is the time axis $\mathbb{R}$, and the group action Φ_t is assumed to preserve some measure on M (for example, this is true for Lebesgue measure on $M = \mathbb{R}^{2d}$, if the vector field $V : M \to \mathbb{R}^{2d}$ is a Hamiltonian vector field).

  In the course of investigation, however, more abstract phase spaces will be used. Our first task is thus to recapitulate the basic notions of measure theory. A good reference for ergodic theory is the book [Wa] by Walters.

Definition 1. *A measurable space is a pair* $(M, \mathcal{M})$, *with a set M and a family $\mathcal{M}$ of subsets of M (the* measurable sets*), such that*

- $M \in \mathcal{M}$
- *If* $A_n \in \mathcal{M}$ $(n \in \mathbb{N})$ *then* $\bigcup_{n\in\mathbb{N}} A_n \in \mathcal{M}$.
- *If* $A \in \mathcal{M}$ *then* $A^c := M \setminus A \in \mathcal{M}$.

$\mathcal{M}$ *is called a σ–algebra on M.*

Example 1. – The power set $\mathcal{P}(M)$ is the largest σ–algebra on a set M.
- The set $\{M, \emptyset\}$ is the smallest σ–algebra.
- If $(M, \mathcal{O})$ is a *topological space* (that is, $\mathcal{O}$ is a family of so-called *open sets* closed under unions and finite intersections, with $M \in \mathcal{O}$ and $\emptyset \in \mathcal{O}$), then there exists a smallest σ–algebra $\mathcal{M}$ of M with $\mathcal{O} \subset \mathcal{M}$, called the σ–algebra of *Borel sets*.
  This is the σ–algebra typically used in applications.

Definition 2. – *A measure on a measurable space* $(M, \mathcal{M})$ *is a map* $\mu : \mathcal{M} \to [0, \infty]$ *(with* $\mu(\mathcal{M}) \neq \{\infty\}$*) which is countably additive, i.e.*

$$\mu\left(\bigcup_{n\in\mathbb{N}} A_n\right) = \sum_{n\in\mathbb{N}} \mu(A_n)$$

if $A_m \cap A_n = \emptyset$ *for* $m \neq n$.
- *μ is a probability measure if in addition* $\mu(M) = 1$.

- A measure space $(M, \mathcal{M}, \mu)$ *is a measurable space* $(M, \mathcal{M})$ *with a measure* $\mu : \mathcal{M} \to [0, \infty]$.
- *If* μ *is a probability measure, then* $(M, \mathcal{M}, \mu)$ *is called a probability space.*

Example 2. 1. If M is a (locally compact) abelian group, then there exists a *translation-invariant*[2] measure μ on a σ–algebra $\mathcal{M}$ containing the Borel sets, called *Haar measure*. This is unique up to normalization and can be chosen to be a probability measure if M is compact.

   In the case of the additive group $M = \mathbb{R}^d$ we obtain *Lebesgue measure*.

2. If h is a regular value of a smooth function $H : \mathbb{R}^d \to \mathbb{R}$, and if the level set $M := H^{-1}(h) \subset \mathbb{R}^d$ is compact (and $\neq \emptyset$), then one constructs a probability measure λ_h on this $(d-1)$–dimensional submanifold by setting

$$\lambda_h(B) := c \cdot \lim_{\varepsilon \searrow 0} \frac{\lambda(\hat{B} \cap H^{-1}((h - \varepsilon, h + \varepsilon)))}{\varepsilon}.$$

   Here λ is Lebesgue measure on $\mathbb{R}^d$. $\hat{B} \subset \mathbb{R}^d$ is the set obtained from B by taking the union of centered straight lines of (small) length $2\delta > 0$ through $b \in B$ and perpendicular to $H^{-1}(h)$.

   Finally, $c := \left(\lim_{\varepsilon \searrow 0} \frac{\lambda(\hat{M} \cap H^{-1}((h-\varepsilon, h+\varepsilon)))}{\varepsilon} \right)^{-1}$ is a normalization constant. The measure λ_h is called *Liouville measure*.

Definition 3. – *A map* $T : M_1 \to M_2$ *between measure spaces* $(M_i, \mathcal{M}_i, \mu_i)$ *is called* measurable *if* $T^{-1}(A_2) \in \mathcal{M}_1$ $(A_2 \in \mathcal{M}_2)$.

- *A measurable map* $T : M_1 \to M_2$ *is called* measure-preserving *if*

$$\mu_1(T^{-1}(A_2)) = \mu_2(A_2) \quad (A_2 \in \mathcal{M}_2).$$

- *A group action of a group* G *on a measurable space* M *is a map*[3]

$$\Phi : G \times M \to M \quad \text{with} \quad \Phi_{\mathrm{Id}} = \mathrm{Id}_M \quad \text{and} \quad \Phi_{g_1} \circ \Phi_{g_2} = \Phi_{g_1 \circ g_2}$$

   for $\Phi_g : M \to M$, $\Phi_g(m) := \Phi(g, m)$.

- *A measure-preserving dynamical system is a quadruple* $(M, \mathcal{M}, \mu, \Phi)$ *consisting of a measure space* $(M, \mathcal{M}, \mu)$ *and a group action* Φ *of an abelian group* G, *such that the maps* $\Phi_g (g \in G)$ *are measure-preserving for* μ.

Example 3. 1. If $H : M \to \mathbb{R}$ is a smooth function on the phase space $M := \mathbb{R}_p^d \times \mathbb{R}_q^d$, then $V := \begin{pmatrix} -\nabla_q H \\ \nabla_p H \end{pmatrix} : M \to \mathbb{R}^{2d}$ is called the *Hamiltonian vector field* of the *Hamilton* function H.

   Assume that it generates a flow

$$\Phi_t : M \to M \quad (t \in \mathbb{R}).$$

   Then the group action $\Phi : \mathbb{R} \times M \to M$, $\Phi(t, m) := \Phi_t(m)$ preserves Lebesgue measure on M.

---

[2] $\mu(A + a) = \mu(A)$ for all $A \in \mathcal{M}$ and $a \in M$

[3] measurable on $(G \times M, \mathcal{G} \times \mathcal{M})$ for a σ-algebra $\mathcal{G}$ of G

2. If, in the same setting, λ_h is Liouville measure on the regular *energy surface* $M := H^{-1}(h)$, then Φ_t restricts to M (since $\frac{d}{dt} H(\Phi_t(m)) \mid_{t=0} = -\nabla_p H(m) \cdot \nabla_q H(m) + \nabla_q H(m) \cdot \nabla_p H(m) = 0$), and $\Phi_t^h := \Phi_t X_M$ is preserving λ_h.

3. The quotient group $\mathbb{R}/\mathbb{Z}$ is isomorphic to the unit circle $S^1 := \{c \in \mathbb{C} \mid |c| = 1\}$ via the isomorphism

$$I : \mathbb{R}/\mathbb{Z} \to S^1 \quad , \quad [x] \mapsto \exp(2\pi i x).$$

Then, given $\alpha \in [0, 1)$, the abelian group $\mathbb{Z}$ acts on $M := \mathbb{R}/\mathbb{Z}$ by the *shifts* $\Phi_n([x]) := [x + n\alpha] \quad (n \in \mathbb{Z}, x \in \mathbb{R})$.

On S^1 this corresponds to a rotation by the multiples of the angle $2\pi\alpha$. By definition of Haar measure μ this group action preserves μ.

4. We consider a fixed $\tilde{T} \in SL(d, \mathbb{Z})$, that is, a $d \times d$ matrix with integer entries and determinant $\det(\tilde{T}) = 1$. Then $\tilde{\Phi}_n : \mathbb{R}^d \to \mathbb{R}^d$ $(n \in \mathbb{Z})$ (defined by $\tilde{\Phi}_0 := \mathrm{Id}$, $\tilde{\Phi}_{k+1} := \tilde{T} \circ \tilde{\Phi}_k$ and $\tilde{\Phi}_{-k-1} := \tilde{T}^{-1} \circ \tilde{\Phi}_{-k}$, $k \geq 0$) is a linear group action of the group $G := \mathbb{Z}$ which has the property

$$\tilde{\Phi}_n(l) \in \mathbb{Z}^d \qquad (l \in \mathbb{Z}^d, n \in \mathbb{Z})$$

to leave the lattice $\mathbb{Z}^d \subset \mathbb{R}^d$ invariant. Thus

$$\tilde{\Phi}_n(x + l) - \tilde{\Phi}_n(x) \in \mathbb{Z}^d \quad \text{for all} \quad x \in \mathbb{R}^d$$

so that $\tilde{\Phi}_n$ descends to a group action

$$\Phi_n : \mathbb{T}^d \to \mathbb{T}^d \qquad (n \in \mathbb{Z})$$

on the d–dimensional torus $\mathbb{T}^d := \mathbb{R}^d/\mathbb{Z}^d \cong (\mathbb{R}/\mathbb{Z})^d$. As the $\tilde{\Phi}$–action preserves Lebesgue measure = Haar measure on $\mathbb{R}^d$, the Φ–action preserves Haar measure μ on the torus $\mathbb{T}^d$.

3 Ergodicity

One way to analyze a group action $\Phi : G \times M \to M$ is to consider the subset

$$\mathcal{I} := \{A \in \mathcal{M} \mid \forall g \in G : \Phi_g(A) = A\}$$

of $\mathcal{M}$ consisting of Φ–invariant sets. As one checks easily, $\mathcal{I}$ is a σ–algebra of M.

Roughly speaking, the more Φ mixes M, the smaller is $\mathcal{I}$. As we are now interested in group actions leaving a probability measure μ invariant, we concentrate on the sizes $\mu(A)$ of the invariant sets $A \in \mathcal{I}$.

Definition 4. *The group action Φ is called* ergodic *w.r.t. the probability measure μ on M if*

$$\mu(A) \in \{0, 1\} \qquad (A \in \mathcal{I}).$$

As $\{\emptyset, M\} \subset \mathcal{I}$, this is the smallest choice of measures occurring.

Example 4. 1. If the energy level $M := H^{-1}(h)$ for the regular value $h \in H(\mathbb{R}^2)$ of the Hamiltonian function $H : \mathbb{R}^2 \to \mathbb{R}$ is compact and connected, then this one-dimensional manifold is known to be diffeomorphic to the circle S^1. In that case there is a smallest *period* $T > 0$ with $\Phi_T^h = \mathrm{Id}_M$.

Moreover, the group $\mathbb{R}$ acts *transitively* on M, that is there is only one *orbit*
$$\mathcal{O}_x := \{\Phi_t^h(x) \mid t \in \mathbb{R}\},$$
independent of $x \in M$.

Thus in that case $\mathcal{I} = \{\emptyset, M\}$ so that the flow is ergodic.

2. Considering the shift action $\Phi_n([x]) = [x + n\alpha]$ on $M = \mathbb{R}/\mathbb{Z}$, one immediately sees that this action of the group $\mathbb{Z}$ cannot be ergodic if $\alpha \in \mathbb{Q}$. Namely if $\alpha = p/q$ with $q \in \mathbb{N}$, $p \in \mathbb{Z}$, then the Borel set

$$A := \bigcup_{k=0}^{q-1} \left[\frac{k}{q}, \frac{k + \frac{1}{2}}{q} \right]$$

belongs to $\mathcal{I}$, but $\mu(A) = 1/2$.

Conversely we would like to prove that the group action of example 4.2 is ergodic if $\alpha \notin \mathbb{Q}$. Here it is useful to apply techniques from functional analysis. For a general measure preserving dynamical system we consider the action

$$\hat{\Phi}_g : L^2(M, \mu) \to L^2(M, \mu) \quad , \quad f \mapsto f \circ \Phi_g \qquad (g \in G).$$

This is a family of unitary operators on the Hilbert space $L^2(M, \mu)$, since by Φ_g-invariance of μ

$$\begin{aligned}
\left\langle \hat{\Phi}_g f_1, \hat{\Phi}_g f_2 \right\rangle &= \int_M f_1 \circ \Phi_g(x) \overline{f}_2 \circ \Phi_g(x) \, d\mu(x) \\
&= \int_M f_1(y) \overline{f}_2(y) \, d\mu(\Phi_{g^{-1}}(y)) \\
&= \int_M f_1(y) \overline{f}_2(y) \, d\mu(y) \\
&= \langle f_1, f_2 \rangle \qquad (f_1, f_2 \in L^2(M, \mu)),
\end{aligned}$$

and $\hat{\Phi}_g$ is onto with inverse $\hat{\Phi}_{g^{-1}}$.

Theorem 1. *The group action Φ is ergodic iff all Φ-invariant $f \in L^2(M, \mu)$ (that is, $\hat{\Phi}_g f = f, g \in G$) are constant μ-almost everywhere.*

Proof:

– Assume that Φ is ergodic and $f \in L^2(M,\mu)$ is Φ–invariant. Then $\mathrm{Re}(f)$ and $\mathrm{Im}(f)$ are $\hat{\Phi}$–invariant, too. So it suffices to assume f to be real. For all $n \in \mathbb{N}$ and $k \in \mathbb{Z}$ the set

$$A_{n,k} := f^{-1}\left([k2^{-n},(k+1)2^{-n})\right)$$

lies in $\mathcal{I}$, and for fixed n we have the partition property

$$A_{n,k_1} \cap A_{n,k_2} = \delta_{k_1,k_2} A_{n,k_1} \quad , \quad \bigcup_{k \in \mathbb{Z}} A_{n,k} = M. \tag{1}$$

By ergodicity of Φ we have $\mu(A_{n,k}) \in \{0,1\}$ and by (1) there exists a unique k_n with $\mu(A_{n,k_n}) = 1$. As $A_{n,k} = A_{n+1,2k}\dot{\cup}A_{n+1,2k+1}$, the sequence $(k_n 2^{-n})_{n \in \mathbb{N}}$ converges to the value attained by f μ–almost everywhere.

– Assume conversely that all $\hat{\Phi}$–invariant $f \in L^2(M,\mu)$ are constant μ–almost everywhere.
  Then if $A \in \mathcal{I}$, the characteristic function $\mathbb{1}_A$ is $\hat{\Phi}$–invariant so that $\mu(A) \in \{0,1\}$. $\qquad\square$

Remark 1. The same argument works for $L^p(M,\mu)$ with arbitrary $p \in [1,\infty]$.

Example 5. If $\alpha \notin \mathbb{Q}$ in Example 3.3, then the shift Φ_n by $n\alpha$ is ergodic. Namely consider a $\hat{\Phi}$–invariant $f \in L^2(\mathbb{R}/\mathbb{Z})$, denote by $e_k \in L^2(\mathbb{R}/\mathbb{Z}), k \in \mathbb{Z}$ the orthonormal basis $e_k([x]) := e^{2\pi ikx}$ and write the Fourier expansion of f in the form

$$f = \sum_{k \in \mathbb{Z}} c_k e_k$$

with coefficients $c_k \in \mathbb{C}$. Then

$$\hat{\Phi}_n f = \sum_{k \in \mathbb{Z}} c_k \hat{\Phi}_n e_k = \sum_{k \in \mathbb{Z}} c_k e^{2\pi ink\alpha} e_k,$$

so that by Φ–invariance for $n = 1$

$$\sum_{k \in \mathbb{Z}} c_k(1 - e^{2\pi ik\alpha})e_k = 0.$$

But since $e^{2\pi ik\alpha} \neq 1$ for $k \neq 0$, $c_k = 0$ for $k \neq 0$ so that f is constant μ–almost everywhere.

4 Mixing

Although group actions of higher-dimensional abelian groups play an important role in statistical mechanics, I confine myself in the sequel to the group

$G = \mathbb{R}$ and $\mathbb{Z}$, that is, flows and invertible iterated maps[4]. As could be seen in the previous examples with one-dimensional phase space, ergodicity is compatible with rather regular and predictable motion. This is not the case with mixing:

Definition 5. *The group action Φ is called* mixing *w.r.t. the probability measure μ on M if*

$$\lim_{|n| \to \infty} \mu(\Phi_n(A) \cap B) = \mu(A) \cdot \mu(B) \qquad (A, B \in \mathcal{M}).$$

Roughly speaking this means that in the limit of infinite time $n \in G$ the set $\Phi_n(A)$ is getting equidistributed on M w.r.t. μ.

Lemma 1. *Mixing implies ergodicity.*

Proof: If $A \in \mathcal{I}$, then mixing implies $\mu(A \cap B) = \mu(A) \cdot \mu(B)$. Setting $B := A$, we get $\mu(A) = (\mu(A))^2$ so that $\mu(A) \in \{0, 1\}$. □

Example 6. We consider the torus automorphism

$$\Phi_n : \mathbb{T}^d \to \mathbb{T}^d \qquad (n \in \mathbb{Z})$$

for the case of a matrix $\tilde{T} \in SL(d, \mathbb{Z})$ whose eigenvalues $\lambda \in \mathbb{C}$ are never *roots of unity* (that is, there is no $l \in \mathbb{N}$ with $\lambda^l = 1$). The simplest example is the so-called *Arnold cat map* with $\tilde{T} = \left(\begin{smallmatrix} 2 & 1 \\ 1 & 1 \end{smallmatrix}\right)$ and $\lambda_{1/2} = \frac{3 \pm \sqrt{5}}{2}$, see Fig. 1.

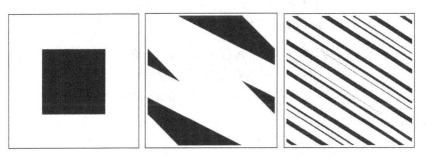

Fig. 1. Action of torus automorphism for matrix $T := \left(\begin{smallmatrix} 2 & 1 \\ 1 & 1 \end{smallmatrix}\right)$. Left: subset $A \subset \mathbb{T}^2$, centre: $\Phi_1(A)$, right: $\Phi_3(A)$.

Denoting the orthonormal Fourier basis of $L^2(\mathbb{T}^d, \mu)$ by $(e_k)_{k \in \mathbb{Z}^d}$, with $e_k(x) := \exp(2\pi i k \cdot x)$, we have

$$\hat{\Phi}_n(e_k)(x) = \exp(2\pi i k \cdot \Phi_n(x)) = \exp\left(2\pi i \tilde{\Phi}_n^t(k) \cdot x\right)$$

$$= e_{\tilde{\Phi}_n^t(k)}(x) \qquad (x \in \mathbb{T}^d).$$

---

[4] Conversely, the thermodynamic formalism of statistical mechanics can be used in the study of dynamical systems, see [Fa] and [17]

As a transposed matrix $\tilde{T}^t$ is in $SL(d, \mathbb{Z})$ if $\tilde{T} \in SL(d, \mathbb{Z})$, $\tilde{\Phi}_n^t$ maps the lattice $\mathbb{Z}^d$ onto itself. As we assumed that no eigenvalue is a root of unity, there is no $k \in \mathbb{Z}^d \setminus \{0\}$ and $n \in \mathbb{Z} \setminus \{0\}$ with

$$\tilde{\Phi}_n^t(k) = k.$$

In other words, given the finite sets

$$\mathbb{Z}_r^d := \{k \in \mathbb{Z}^d \mid |k| \leq r\}$$

for an arbitrary r, all $k \in \mathbb{Z}_r^d \setminus \{0\}$ eventually leave this set, i.e. $\tilde{\Phi}_n^t(k) \notin \mathbb{Z}_r^d$ for $|n| \geq N(r)$. Now, in order to prove mixing, we consider the $L^2(\mathbb{T}^d, \mu)$–functions $f := \mathbb{1}_A$ and $g := \mathbb{1}_B$ and want to prove that

$$\lim_{n \to \infty} (\hat{\Phi}_n f, g) = (f, \mathbb{1}_{\mathbb{T}^d}) \cdot (g, \mathbb{1}_{\mathbb{T}^d}).$$

Given $\varepsilon > 0$ and the Fourier expansions $f = \sum_{n \in \mathbb{Z}^d} c_n e_n$ and $g = \sum_{n \in \mathbb{Z}^d} d_n e_n$ (with $(c_n)_{n \in \mathbb{Z}^d}, (d_n)_{n \in \mathbb{Z}^d} \in l^2(\mathbb{Z}^d)$), there is an $r > 0$ with $\|f - f_r\|_2 < \varepsilon$ and $\|g - g_r\|_2 < \varepsilon$ for $f_r := \sum_{n \in \mathbb{Z}_r^d} c_n e_n$ and $g_r := \sum_{n \in \mathbb{Z}_r^d} d_n e_n$.

Now for $|n| > N(r)$ we have

$$(\hat{\Phi}_n f_r, g_r) = \sum_{k,l \in \mathbb{Z}_r^d} c_k \overline{d}_l \cdot (\hat{\Phi}_n e_k, e_l) = c_0 \overline{d}_0 = (f_r, e_0) \cdot (e_0, g_r)$$

since for $k \in \mathbb{Z}_r^d \setminus \{0\}$ $\hat{\Phi}_n e_k = e_{\tilde{\Phi}_n^t(k)}$ is orthonormal to all the e_l. On the other hand, using unitarity of $\hat{\Phi}_n$,

$$
\begin{aligned}
|(\hat{\Phi}_n f, g) - (\hat{\Phi}_n f_r, g_r)| &\leq |(\hat{\Phi}_n(f - f_r), g)| + |(\hat{\Phi}_n f_r, g - g_r)| \\
&\leq \|(\hat{\Phi}_n(f - f_r)\|_2 \|g\|_2 + \|\hat{\Phi}_n f_r\|_2 \|g - g_r\|_2 \\
&\leq \varepsilon(\|g\|_2 + \|f_r\|_2) \leq \varepsilon(\|g\|_2 + \|f\|_2 + \varepsilon)
\end{aligned}
$$

so that

$$|(\hat{\Phi}_n f, g) - (f, e_0) \cdot (e_0, g)| \leq \varepsilon(\|g\|_2 + \|f\|_2 + \varepsilon)$$

showing the mixing property.

5 The Birkhoff Ergodic Theorem

Ergodic theory of dynamical systems is more than the theory of ergodic dynamical systems. It is also able to tell us something about the long-term behavior of arbitrary measure-preserving dynamical systems, ergodic or not. We may even include situations where the measure μ is not finite and thus cannot be normalized to be a probability measure. We only have to impose the mild assumption that μ is σ–finite.

Definition 6. *A measure space $(M, \mathcal{M}, \mu)$ is called σ-finite if there exists $M_n \in \mathcal{M}$ with $\mu(M_n) < \infty$ and*

$$M = \bigcup_{n \in \mathbb{N}} M_n.$$

Example 7. $M := \mathbb{R}^d$ with Lebesgue measure μ and the balls $M_n := \{x \in \mathbb{R}^d \mid \|x\| \leq n\}$.

Theorem 2 (Birkhoff Ergodic Theorem). *Let $(M, \mathcal{M}, \mu, \Phi)$ be a dynamical system (with group $\mathbb{Z}$) on a σ-finite measure space and $f \in L^1(M, \mu)$.*
- *Then $\frac{1}{n} \sum_{k=0}^{n-1} \hat{\Phi}_k f$ converges μ-almost everywhere to a function $f^+ \in L^1(M, \mu)$, with $\hat{\Phi}(f^+) = f^+$ (μ-almost everywhere).*
- *If $\int_M d\mu < \infty$, then $\int_M f \, d\mu = \int_M f^+ \, d\mu$.*

We omit the proof which can be found in [Wa]. A similar statement holds for $G = \mathbb{R}$. See [KH] for another proof in the case of a probability measure and [17] for a generalization to $\mathbb{Z}^d$-actions.

Whereas in the theorem the positive time limit was considered, we may equally well consider the negative time limit (in the L^1-sense)

$$f^- := \lim_{n \to \infty} \frac{1}{n} \sum_{k=0}^{n-1} \hat{\Phi}_{-k} f.$$

Corollary 1. *If $\int_M d\mu < \infty$, then $f^+ = f^-$ μ-almost everywhere.*

Proof: We may consider the real and imaginary part of f separately and thus assume f to be real. Then for $\varepsilon > 0$ set

$$A_\varepsilon := \{x \in M \mid f^+(x) \geq f^-(x) + \varepsilon\}. \tag{2}$$

We consider now $g := f \cdot \mathbb{1}_{A_\varepsilon} \in L^1(M, \mu)$. Applying the Birkhoff Ergodic Theorem to g, we get $\int_{A_\varepsilon} f \, d\mu = \int_M g \, d\mu = \int_M g^\pm \, d\mu$.

Moreover, $A_\varepsilon \in \mathcal{I}$ so that[5] $\int_M g^\pm \, d\mu = \int_{A_\varepsilon} f^\pm \, d\mu$. Thus

$$0 = \int_{A_\varepsilon} (f^+ - f^-) \, d\mu \geq \varepsilon \mu(A_\varepsilon),$$

i.e. $\mu(A_\varepsilon) = 0$ for all $\varepsilon > 0$. This implies $f^+(x) \leq f^-(x)$ μ-almost everywhere. Exchanging f^+ and f^- in (2) gives the converse inequality. $\qquad \square$

Metaphorically speaking this tells us that the past equals the future (but only in L^1-sense!). This is a highly non-trivial statement.

---

[5] To be precise, as $f^\pm \in L^1(M, \mu)$, we have

$$A_\varepsilon \in \mathcal{I}_\mu := \{A \in \mathcal{M} \mid \forall g \in G : \mu(\Phi_g(A) \triangle A) = 0\} \supset \mathcal{I}$$

Example 8. For the Hamilton function $H(p,q) := \frac{1}{2}p^2 + V(q)$ on $\mathbb{R}_p^d \times \mathbb{R}_q^d$ the Hamilton equation can be rewritten in Newton's form $\ddot{q} = -\nabla V(q)$ with the force term $-\nabla V$.

Assuming that $V \in C^2(\mathbb{R}^d, \mathbb{R})$ is bounded below by $V_{\min}$, for energy E the speed $|p|$ is bounded above by $|p| \leq \sqrt{2(E - V_{\min})}$. This means that the flow Φ_t exists for all $t \in \mathbb{R}$.

Now consider two cases:

1. V is $\mathcal{L}$–periodic w.r.t. a regular lattice $\mathcal{L} := \{n_1 l_1 + \ldots + n_d l_d \mid n_i \in \mathbb{Z}\}$ (with $l_1, \ldots, l_d$ a basis of $\mathbb{R}^d$). Then we may regard V as a function $\hat{V}$ on the d–dimensional torus $\mathbb{T} := \mathbb{R}^d / \mathcal{L}$, and H as the function

$$\hat{H} : \mathbb{R}_p^d \times \mathbb{T} \to \mathbb{R} \quad , \quad \hat{H}(p,q) := \frac{1}{2}p^2 + \hat{V}(q).$$

Unlike for H, the energy surfaces $\hat{H}^{-1}(E)$ are compact. Consider (for a regular value E of $\hat{H}$) Liouville measure $\hat{\lambda}_E$ on $\hat{H}^{-1}(E)$. Then the means

$$\bar{v}^+(\hat{x}) := \lim_{T \to \infty} \frac{1}{T} \int_0^T p(t, \hat{x}) \, dt$$

and

$$\bar{v}^-(\hat{x}) := \lim_{T \to \infty} \frac{1}{T} \int_{-T}^0 p(t, \hat{x}) \, dt$$

exist for $\hat{\lambda}_E$–almost every initial point $\hat{x} \in \hat{H}^{-1}(E)$ and are equal, according to Birkhoff's ergodic theorem and its corollary 1.
If a phase space point $x \equiv (p_0, q_0) \in H^{-1}(E)$ projects to $\hat{x} \in \hat{H}^{-1}(E)$, then the limit

$$\bar{v}^\pm(x) := \lim_{T \to \infty} \frac{1}{T}(q(T, x) - q_0)$$

exists and equals $\bar{v}^\pm(\hat{x})$. Thus this *asymptotic (mean) velocity* exists λ_E–almost everywhere w.r.t. the Liouville measure λ_E on the non-compact energy shell $H^{-1}(E)$

2. Now let $V \in C^2(\mathbb{R}^d, \mathbb{R})$ with $\lim_{|q| \to \infty} V(q) = 0$, i.e. a scattering situation.
   If we assume in addition a fast enough fall-off of V and its derivatives (say, V being of compact support for simplicity), then the phase space functions $\bar{v}^+$ and $\bar{v}^-$ exist on $H^{-1}(E)$ for any energy $E > 0$.
   Here, however, $\bar{v}^+$ is typically unequal to $\bar{v}^-$ on a set of positive Liouville measure. In fact, for scattering orbits the speeds $|\bar{v}^\pm(x)|$ are both equal to $\sqrt{2E}$ by energy conservation, whereas the change in directions $\bar{v}^\pm(x)/\sqrt{2E}$ is the subject of the theory of classical potential scattering.

From now on we assume $\int_M d\mu < \infty$.

Corollary 2. *If the dynamical system (with group $\mathbb{Z}$) is ergodic, then for $f \in L^1(M, \mu)$*

$$f^\pm \in L^1(M, \mu)$$

are constant μ–almost everywhere.

Proof: This follows from Remark 1. □

In other words, for μ–almost every $x \in M$ the *time means* $f^{\pm}(x)$ coincide with the *space mean* $\int_M f \, d\mu$.

Example 9. Consider the torus automorphism Φ generated by $\tilde{T} = \left(\begin{smallmatrix} 2 & 1 \\ 1 & 1 \end{smallmatrix} \right) \in SL(2, \mathbb{Z})$, and for $n \in \mathbb{N}$ the projection $L_n \subset \mathbb{T}^2$ of the lattice

$$\tilde{L}_n := \left\{ \left(\begin{smallmatrix} k_1/n \\ k_2/n \end{smallmatrix} \right) \mid k_1, k_2 \in \mathbb{Z} \right\} \subset \mathbb{R}^2.$$

This lattice L_n is permuted by Φ. It consists of n^2 points, and for a given continuous function $f : \mathbb{T}^2 \to \mathbb{R}$ the restriction $f^+ X_{L_n}$ has constant values on the cycles of these permutations. These values, however, in general differ from cycle to cycle, although T is ergodic (even mixing). One notices that $\bigcup_{n \in \mathbb{N}} L_n$ is dense in $\mathbb{T}^2$. This example shows that one cannot in general omit the restrictive "μ–almost everywhere" in the last theorem.

Finally, we compare the notions of ergodicity and of mixing. We recall the mixing property (Def. 5)

$$\lim_{|k| \to \infty} \mu(\Phi_k(A) \cap B) = \mu(A) \cdot \mu(B)$$

of iterated maps.

Corollary 3. *The group action Φ is ergodic w.r.t. the probability measure μ on M iff*

$$\lim_{n \to \infty} \frac{1}{n} \sum_{k=0}^{n-1} \mu(\Phi_k(A) \cap B) = \mu(A) \cdot \mu(B). \tag{3}$$

Proof: • Set $f_n := \frac{1}{n} \sum_{k=0}^{n-1} \mathbb{1}_{\Phi_k(A)} \in L^1(M, \mu)$.

According to Theorem 2 these functions converge μ–almost everywhere to a $\hat{\Phi}$–invariant function $f^+ \in L^1(M, \mu)$. According to Remark 1 f^+ is constant μ–almost everywhere if Φ is ergodic, and

$$\int_M f^+ \, d\mu = \int_M \mathbb{1}_A \, d\mu = \mu(A).$$

So

$$\lim_{n \to \infty} \frac{1}{n} \sum_{n=0}^{n-1} \mu(\Phi_k(A) \cap B) = \lim_{n \to \infty} \int f_n \cdot \mathbb{1}_B \, d\mu = \mu(A) \cdot \mu(B).$$

• Conversely for the case $A = B \in \mathcal{I}$ we have $\Phi_k(A) \cap B = A$ so that, assuming (3), we conclude $\mu(A) = \mu(A)^2$, hence ergodicity. □

6 Hyperbolicity

In our basic examples of ergodicity and mixing we profited from the fact that the phase space of our system was an abelian group (in fact a torus).

This, however, is somewhat untypical in applications of ergodic theory. In particular the typical physical problem we have in mind is that of the hamiltonian flow, which has the energy surface as its phase space. So now we consider geometric properties of an iterated map or a flow which can be used to prove the mixing property.

Definition 7. *Let M be a manifold.*

- *Then a distribution on M is a subbundle of the tangent bundle TM of M.*
- *A distribution E is called integrable, if for any vector fields[6] $X, Y : M {\to} E$ the commutator vector field $[X, Y] : M \to TM$ only takes values in E.*

Example 10. 1. If $X : M \to TM$ is a (nonvanishing) vector field and $E(m) := \operatorname{span}(X(m)) \subset T_m M$, then E is an integrable distribution.
2. A nonvanishing one-form ω on M has a null-space of dimension $\dim(M) - 1$ and thus defines a distribution E on M. Then the formula

$$\omega([X, Y]) = d\omega(X, Y) - X\omega(Y) + Y\omega(X)$$

(see [AM], Chapt. 2.4) gives for vector fields $X, Y : M \to E$ the criterion $d\omega(X, Y) = 0$ (since then $\omega(X) = \omega(Y) = 0$).
E.g. if on $\mathbb{R}^3$ we have $\omega = x_1 dx_2 + dx_3$, then $d\omega = dx_1 \wedge dx_2$, and ω vanishes on span $\left(\begin{pmatrix} 1 \\ 0 \\ 0 \end{pmatrix}, \begin{pmatrix} 0 \\ 1 \\ 0 \end{pmatrix} \right) \subset T_0 \mathbb{R}^3$, but $d\omega$ does not vanish on this pair of tangent vectors. So the distribution defined by ω is not integrable.

We will encounter distributions related to foliations of phase space.

Definition 8. *A foliation W of class C^r and dimension k on the m-dimensional manifold M is a partition of M into connected, immersed C^r-manifolds[7] $W(x)$ through $x \in M$ (called the leaves of the foliation), for which each point x of M has a neighborhood U and coordinates $(y, z) : U \to \mathbb{R}^k \times \mathbb{R}^{m-k}$ such that the connected component W_U of x within $W(x) \cap U$ is given by the equation*

$$W_U(x) = z^{-1}(c) \quad with \quad c := z(x).$$

Recall that $f : N \to M$ is called an *immersion* if its linearizations $T_n f : T_n N \to T_{f(n)} M$ $(n \in N)$ are injective.

Example 11. 1. If $H \in C^\infty(M, \mathbb{R})$ is a hamiltonian function on a $(2d)$-dimensional symplectic manifold M, then it is called *integrable* if there are functions $f_1, \ldots, f_d \in C^\infty(M, \mathbb{R})$ with $f_1 = H$ which are independent

---

[6] More generally, one allows X, Y to be defined on an open subset $U \subset M$.
[7] See, e.g. [AM], Chap. 1.6

and Poisson-commute ($\{f_i, f_j\} = 0$ for $i, j = 1, \ldots, d$). Then the connected components of the level sets $F^{-1}(c)$ with $F := \begin{pmatrix} f_1 \\ \vdots \\ f_d \end{pmatrix} : M \to \mathbb{R}^d$ of $c \in F(M)$ form leaves of a C^∞–foliation of M of dimension d (see, e.g. [AM], Chapt. 5.2).

2. For $v \in \mathbb{R}^d \setminus \{0\}$ the affine subspaces

$$W(x) := \{x + cv \mid c \in \mathbb{R}\}$$

through $x \in \mathbb{R}^d$ form the leaves of a one-dimensional C^∞–foliation of $\mathbb{R}^d$. Their images under the projection $\mathbb{R}^d \to \mathbb{R}^d / \mathbb{Z}^d$ on the torus $\mathbb{T}^d := \mathbb{R}^d / \mathbb{Z}^d$ form the leaves of a one-dimensional C^∞–foliation of $\mathbb{T}^d$. If $v = \begin{pmatrix} \alpha \\ 1 \end{pmatrix}$ with $\alpha \in \mathbb{R} \setminus \mathbb{Q}$, then these leaves are all dense in $\mathbb{T}^2$.

Definition 9. *Let* $\Phi : G \times M \to M$ *be a smooth group action of* $G = \mathbb{R}$ *or* $\mathbb{Z}$ *on a manifold* M *with metric* d.

Then the stable *respectively* unstable manifold *of* $x \in M$ *is the set*

$$V^s(x) := \{y \in M \mid \lim_{t \to \infty} d(\Phi_t(y), \Phi_t(x)) = 0\}$$

respectively

$$V^u(x) := \{y \in M \mid \lim_{t \to -\infty} d(\Phi_t(y), \Phi_t(x)) = 0\}.$$

M is partitioned into (un-)stable manifolds.

Example 12. For the Arnold cat map of Example 6 the stable manifold $V^s([x])$ of $[x] \in \mathbb{T}^2$, $x \in \mathbb{R}^2$ is the torus projection of the affine space $\{x + cv_2 \mid c \in \mathbb{R}\}$ and $V^u([x])$ the one of $\{x \in cv_1 \mid c \in \mathbb{R}\}$ where $v_i = \begin{pmatrix} \alpha_i \\ 1 \end{pmatrix}$ is an eigenvector for the eigenvalue λ_i.

As $\alpha_{1/2} = \frac{1}{2}(1 \pm \sqrt{5}) \notin \mathbb{Q}$, the stable and unstable manifolds are dense in $\mathbb{T}^2$.

In general $V^s(x)$ and $V^u(x)$ may be just measurable sets. Because of the following theorem they are, however, interesting in the analysis of ergodic properties of Φ.

Theorem 3. *Let* Φ *be a continuous group action on a compact manifold* M *preserving a probability measure* μ, *where* $\mu(U) > 0$ *for* $U \neq \emptyset$ *open.*

Then a $\hat{\Phi}$–invariant $f \in L^1(M, \mu)$ *is constant on the (un-) stable manifolds, up to sets of measure zero.*

Proof:

– Consider first a continuous function g on M, not necessarily $\hat{\Phi}$–invariant. By compactness of M it is *uniformly continuous*, i.e. for $\varepsilon > 0$ there is a $\delta > 0$ with

$$|g(x) - g(y)| < \varepsilon \quad \text{for} \quad d(x,y) < \delta.$$

Here d is an arbitrarily chosen metric on M. Hence for $y \in V^s(x)$

$$\lim_{n \to \infty} \frac{1}{n} \sum_{k=0}^{n-1} (g(\Phi_k(x)) - g(\Phi_k(y))) = 0,$$

so that $g^+(y) = g^+(x)$ if $g^+(x)$ exists, that is, up to a set $N^s(g)$ of measure zero.

- We now approximate $f \in L^1(M, \mu)$ in the L^1–sense by continuous functions $(g_m)_{m \in \mathbb{N}}$. Then by Φ–invariance of f and μ we get

$$\int_M \left| f - \frac{1}{n} \sum_{k=0}^{n-1} g_m \circ \Phi_k \right| d\mu \le \frac{1}{n} \sum_{k=0}^{n-1} \int_M |f \circ \Phi_k - g_m \circ \Phi_k| \, d\mu$$

$$= \frac{1}{n} \sum_{k=0}^{n-1} \int_M |f - g_m| \, d\mu = \int_M |f - g_m| \, d\mu,$$

that is

$$\int_M |f - g_m^+| \, d\mu \le \int_M |f - g_m| \, d\mu.$$

So f is the L^1–limit of the g_m^+. Thus by going to a subsequence there is a measure zero set $N^s(f)$ with the pointwise limit

$$\lim_{m \to \infty} g_m^+(x) = f(x) \qquad (x \in M \setminus N^s(f)).$$

(Here we use the positivity of μ on open sets.)

- The proof is similar for the unstable manifolds and for case of the group $G = \mathbb{R}$. $\square$

Remark 2. We now see how in the example of the cat map Theorem 3 is the key to an alternative proof of ergodicity. For a small open disk U centered at $x \in \mathbb{T}^2$ one takes the connected component $V_U^s(x)$ of $V^s(x) \cap U$ containing x. For typical $x \in \mathbb{T}^2$, that is those x not belonging to $N^s(f)$ for the function $f \in L^1(\mathbb{T}^2, \mu)$ under consideration, f^+ is constant on $V_U^s(x)$. Having chosen such an x, for typical $y \in V_U^s(x)$ the limit function f^+ is constant on $V_U^u(y)$ (otherwise one could construct a positive measure subset of $z \in U$ on which f^+ is not defined or not constant along the leaves $V^u(z)$). Thus almost everywhere on U the function f^+ is constant. Taking an open covering of $\mathbb{T}^2$ by such sets $U_1, \ldots, U_n$, we see that the Arnold cat map meets the condition for ergodicity.

The above proof of ergodicity has the merit of being based on the geometric structure of the stable and unstable manifolds (that is, subsets of phase space defined by the dynamics) and not using the fact that $\mathbb{T}^2$ is an abelian group. This opens the door to a proof working for many hyperbolic dynamical systems.

We formulate the results in terms of *Riemannian manifolds* (M, g), that is, a manifold M with a scalar product g_x on each tangent space $T_x M$, smoothly depending on $x \in M$.

The *length* $(g_x(Y, Y))^{1/2}$ of a tangent vector $Y \in T_x M$ is then denoted by $\|Y\|$.

Let Φ be a smooth dynamical system with group $G = \mathbb{R}$ or $\mathbb{Z}$ on a compact Riemannian manifold (M, g). In the case $G = \mathbb{R}$ we assume the vector field $X := \frac{d}{dt} \Phi_t \mid_{t=0}$ to be non-vanishing, and denote by $E^0(x)$ the subspace span$(X(x))$ of $T_x M$.

Definition 10. Φ *is called* Anosov, *if there are* Φ–*invariant distributions* $E^s, E^u \subset TM$ *and constants* $c > 0, \lambda \in (0, 1)$ *such that for all* $x \in M$

a) $E^s(x) \oplus E^u(x) = T_x M$ *if* $G = \mathbb{Z}$,
   $E^s(x) \oplus E^u(x) \oplus E^0(x) = T_x M$ *if* $G = \mathbb{R}$.
b) $\|d\Phi_t(x)(v)\| \leq c\lambda^{\pm t}\|v\|$ $(t \geq 0, v \in E^{s/u}(x))$.

Example 13. Setting $E^s(x) := T_x V^s(x)$ and $E^u(x) := T_x V^u(x)$ for the Arnold cat map, we see that this dynamical system is ergodic (we may choose $c := 1$ and $\lambda := \lambda_2$ if the metric g is the torus projection of the Euclidean metric on $\mathbb{R}^2$).

According to the Frobenius Theorem, a distribution is integrable if and only if it is the distribution of tangent spaces of a regular foliation. In our case the leaves for the distributions $E^{s/u}(x)$ are the (un-)stable manifolds $V^{s/u}(x)$.

The most important example of an Anosov flow is the one of the geodesic flow on a compact Riemannian manifold of negative sectional curvature.

7 Geodesic Flows

Up to now we used the Riemannian metric g on a d–dimensional manifold N only to measure the length of tangent vectors. Now we use it to define the geodesic flow on the tangent bundle TN of N. For simplicity we assume that N is a *closed* manifold, that is, a compact manifold without boundary. Then for a curve $c \in C^1(I, N)$ defined on an interval $I = [a, b]$ we call

$$\mathcal{E}(c) := \tfrac{1}{2} \int_a^b g_{c(t)}(\dot{c}(t), \dot{c}(t))\, dt$$

the *energy* of c. The critical points of the functional $\mathcal{E}$ meet the Euler-Langrange-equation

$$\frac{d}{dt} \frac{\partial L}{\partial v}(c, \dot{c}) - \frac{\partial L}{\partial q}(c, \dot{c}) = 0$$

for the Langrange function $L(q,v) := \frac{1}{2}g_q(v,v)$. In a local coordinate system the metric has the form $g_x(X,Y) = \sum_{i,k=1}^{d} g_{i,k}(x)X_iY_k$, and the Euler-Lagrange equation is the *geodesic equation*

$$\ddot{c}_i + \Gamma^i_{j,k}(c)\dot{c}_j\dot{c}_k = 0$$

with the *Christoffel symbol*

$$\Gamma^i_{jk} = \frac{1}{2}g^{il} \cdot \left(\frac{\partial g_{jl}}{\partial q_k} + \frac{\partial g_{lk}}{\partial q_j} - \frac{\partial g_{jk}}{\partial q_l} \right).$$

Here the Einstein convention to sum over pairs of identical indices is employed, and $g^{ir}(q)g_{rs}(q) = \delta_{is}$, that is, the matrix $(g^{ir}(q))$ is inverse to $(g_{rs}(q))$.

The *covariant derivative* $\nabla_X Y$ of the vector field Y in the direction of the vector field X on (N,g) is locally given by its components

$$(\nabla_X Y)_i = \frac{\partial Y_i}{\partial x^j}X_j + \Gamma^i_{j,k}X_jY_k \qquad (i = 1\ldots, \dim(N)). \tag{4}$$

So the geodesic equation can be written[8] as $\nabla_{\dot{c}}\dot{c} = 0$.

One verifies by a direct calculation that

$$\frac{d}{dt}g_{c(t)}(\dot{c}(t), \dot{c}(t)) = 0.$$

In other words, the speed $\sqrt{g_c(\dot{c}, \dot{c})} = \|\dot{c}\|$ of the geodesic is constant. This immediately implies that the geodesics, defined as extremizers of $\int \|\dot{c}\|^2\, dt$, extremize the length functional, $\int \|\dot{c}\|\, dt$, too and are thus locally shortest curves between their end points. Secondly, we may normalize the speed to equal one and restrict the geodesic flow to the so-called *unit tangent bundle*

$$SN := \{X \in TN \mid \|X\| = 1\}.$$

Henceforth we denote this closed $(2d - 1)$–dimensional manifold by M and the geodesic flow on it by

$$\Phi_t : M \to M \qquad (t \in \mathbb{R}).$$

By compactness of M it exists for all times, and it leaves the Liouville measure μ on M invariant, since $L : TN \to \mathbb{R}$ can be seen to be the Hamiltonian function of the geodesic flow.[9] Advanced texts on geodesic flows are the books [Kl] by W. Klingenberg and [Pa] by G. Paternain.

---

[8] extending the velocity $\dot{c}$ along the geodesic arbitrarily

[9] With respect to the symplectic form on TN which arises from the canonical symplectic form on T^*N by the isomorphism given by g.

Although the form of the metric tensor g depends on the coordinate system, there are well-known geometric invariants of the Riemannian manifold (N, g) encoded by the *Riemann curvature tensor* R defined by

$$R(X, Y)Z := \nabla_Y \nabla_X Z - \nabla_X \nabla_Y Z + \nabla_{[X,Y]} Z$$

for vector fields $X, Y, Z : N \to TN$. As the coordinate representation (4) of the covariant derivative ∇_X involves the Christoffel symbol, the one of the curvature tensor involves the metric tensor and its first two derivatives.

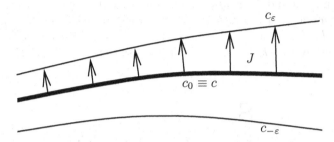

Fig. 2. Jacobi vector field J along a geodesic c

Now as the geodesic equation contains the covariant derivative $\nabla_{\dot{c}}$, it should not come as a surprise that the *Jacobi equation*, its linearization, contains R:

$$\ddot{J} + R_c(\dot{c}, J)\dot{c} = 0$$

Here J is a vector field along c arising from the linearization of the geodesic flow, that is, a one parameter family c_ε of geodesics, $|\varepsilon| < \varepsilon_0$:

$$J = \frac{d}{d\varepsilon} c_\varepsilon \big|_{\varepsilon=0},$$

see Fig. 2. The operators

$$K(t) := R_{c(t)}(\dot{c}(t), \cdot)\dot{c}(t) \qquad (t \in R)$$

on $T_{c(t)}N$ are symmetric, and negative definite after reduction to the subspace of vectors perpendicular to $\dot{c}(t)$, if the metric is of negative curvature. We thus study the linear ordinary differential equation[10]

$$\ddot{J}(t) + K(t)J(t) = 0. \tag{5}$$

We now consider a $(d-1)$–dimensional subspace of $T_x M$ of initial conditions $J(0), \dot{J}(0)$ of the Jacobi equation, and assume the existence of a linear operator $S(0)$ with $\dot{J}(0) = S(0)J(0)$. Then at least in a small time interval

---

[10] however, we should keep in mind that the time derivatives are covariant derivatives

including zero there exist similar operators $S(t)$ with $\dot{J}(t) = S(t)J(t)$, and we get

$$\ddot{J}(t) = \dot{S}(t)J(t) + S(t)\dot{J}(t) = \left(\dot{S}(t) + S(t)^2\right)J(t).$$

Thus we deduce from the Jacobi equation the *Riccati equation*

$$\dot{S}(t) + S(t)^2 + K(t) = 0.$$

The virtue of this non-linear equation, compared to (5), is that it is of first order. To understand it, we assume for simplicity that N is a surface ($d = 2$). Then S is a scalar equation. Assuming negative Gaussian curvature, we have $K(t) \leq -k_u$ for some $k_u > 0$ so that $\dot{S}(t) \geq 0$ if $|S(t)| \leq \sqrt{k_u}$, but by compactness $K(t) \geq k_l$ for some $k_l \geq k_u > 0$ so that $\dot{S}(t) \leq 0$ if $|S(t)| \geq \sqrt{k_l}$, see Fig. 3.

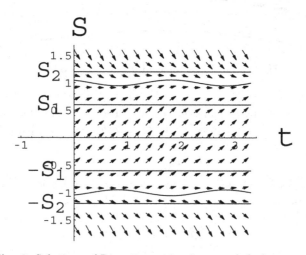

Fig. 3. Solutions of Riccati equation in extended phase space

If $[S_1, S_2]$ is an interval containing the interval $[\sqrt{k_u}, \sqrt{k_l}]$ in its interior, then this is mapped by the forward solution of the Riccati equation into the interior of the interval $[S_1, S_2]$. As $\dot{J}(t) = S(t)J(t)$, that interval of the phase space of the Riccati equation corresponds to a *cone* in the phase space of the Jacobi equation, and these cones are mapped by the forward flow into each other. By a limit argument the intersection of all those cones in the tangent space $T_x M$ consists of a one-dimensional subspace, the unstable subspace, see Fig. 4.

Starting instead with the interval $[-S_2, -S_1]$ and taking the intersections of the corresponding cones in the negative time direction, we get the stable subspace, which is again one-dimensional. A generalization to higher dimension gives $(d - 1)$-dimensional (un-)stable subspaces.

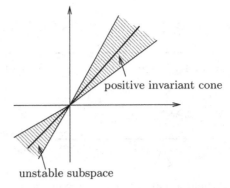

positive invariant cone

unstable subspace

Fig. 4. Positive invariant cone field in the tangent space of phase space

By continuity of the flow these subspaces of $T_x M$ depend continuously on $x \in M$. One thus proves

Theorem 4. *If (N, g) is a closed Riemannian manifold of negative curvature, then the geodesic flow Φ_t on its unit tangent bundle $M = SN$ is an Anosov flow.*

The cone field technique sketched above is useful in showing the existence of stable and unstable distributions, even in cases where the flow is not smooth[11]. The details are described nicely in [27].

Remark 3. For the geodesic motion on a negatively curved suface (N, g) one best visualizes the Anosov structure by looking locally at the stable and unstable manifolds $V^s(x)$ and $V^u(x)$ of $x \in M = SN$.

8 Ergodicity of the Geodesic Flow

Now it *seems* that we are ready to prove ergodicity of the above geodesic flow (assuming N to be connected). We have foliations of M into the stable respectively unstable manifolds $V^s(x)$, $V^u(x)$ which are tangential to the stable respectively unstable distribution just constructed and we have the foliation given by the flow lines $V^0(x) := \Phi(\mathbb{R}, x)$ through $x \in M$, see Fig. 5.
There are, however, two problems left:

- The distribution given by the direct sum $E^s(x) \oplus E^u(x)$ is not integrable.
- In general (for non-constant curvature) the distributions $E^s(x)$ and $E^u(x)$ are not even differentiable in their dependence on $x \in M$.

Remarks 5 *1. The lack of integrability of the distribution $E^s \oplus E^u$ can be seen for the case of a negatively curved surface (N, g) by the following consideration: Suppose we want to synchronize the geodesics with the*

---

[11] This happens, for example for billiards

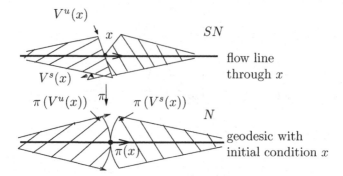

Fig. 5. Stable manifold $V^s(x)$ and unstable manifold $V^u(x)$ of $x \in SN$ (*above*), and their projections to the configuration space N (*below*)

nearby initial conditions $x, y \in U \subset SN$. We can do this in two natural ways starting from x. There is a unique point $z_1 \in V_U^s(x)$ and $y_1 = \Phi_{t_1}(y)$ such that $z_1 \in V_U^u(y_1)$. Similarly, there is a unique point $z_2 \in V_U^u(x)$ and $y_2 = \Phi_{t_2}(y)$ such that $z_2 \in V_U^s(y_2)$. However in general $y_1 \neq y_2$, see Fig. 6.

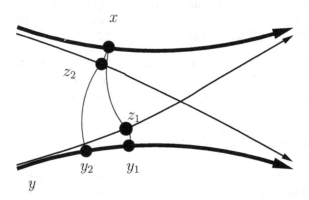

Fig. 6. Nonintegrability of $E^s \oplus E^u$ for the geodesic flow

This lack of integrability is a technical nuisance. On the other hand it is used and needed *in the proof of mixing of the geodesic flow.*

2. *The lack of smoothness of the distributions E^s and E^u can be motivated by the same picture. Consider the unstable distributions $E^u(\Phi_t(x))$ and $E^u(\Phi_t(z_1))$, first for $t = 0$. These distributions depend on the curvatures along the negative time segments of the geodesics through x respectively z_1, which may be very different.*

   In general the stable and unstable foliations of a smooth Anosov flow or map are not even C^1 (although the leaves are smooth). This leads *to problems in generalizing the proof of ergodicity we gave for the torus automorphism. As shown by Peano, there are continuous maps from an interval onto a square, so that sets of Lebesgue measure zero in $\mathbb{R}^2$ can be mapped onto sets of positive measure. As the above mentioned proof used the foliation of the torus by (un-)stable manifolds, if implicitly used the Fubini theorem of integration. In order to apply that argument to the general case, one needs to show* Hölder *continuity of the distributions, that is the existence of a* Hölder *exponent $\alpha > 0$ and a $c > 0$ with*

$$\mathrm{dist}(E^s(x), E^s(y)) \leq c(d(x,y))^{\alpha} \qquad (x,y \in M),$$

and similarly for E^u, with a suitably defined distance function dist. *The Appendix by Misha Brin in the book [Ba] explains how this can be done.*

9 Motion in Periodic Potentials

We have seen that the Birkhoff Ergodic Theorem implies the existence and equality of the asymptotic velocities $\overline{v}^{\pm}(x)$ for the motion in a $\mathcal{L}$–periodic potential $V \in C^2(\mathbb{R}^d, \mathbb{R})$, up to measure zero. So for a given regular value E of V we get a probability distribution ν_E of asymptotic velocities on $\mathbb{R}^d$, which is the image measure of Liouville measure λ_E w.r.t. $\overline{v}^{\pm}$.

What does that mean for the semiclassical limit $\hbar > 0$ of the Schrödinger equation

$$H_{\hbar} := -\frac{\hbar^2}{2}\Delta + V \quad \text{on} \quad L^2(\mathbb{R}^d)?$$

It is well-known that $H_{\hbar}$ is unitary equivalent to the direct integral

$$\int_{\mathbb{T}^*}^{\oplus} H_{\hbar}(k)dk$$

of Schrödinger operators $H_{\hbar}(k) := \frac{1}{2}\left(\frac{\hbar}{i}\nabla + k\right)^2 + \hat{V}$ on $L^2(\mathbb{T})$, where the so-called quasimomentum k varies in the torus $\mathbb{T}^* := \mathbb{R}^d/\mathcal{L}^*$, $\mathcal{L}^*$ being the dual lattice of $\mathcal{L}$. Given the solutions

$$H_{\hbar}(k)\Psi_{j,\hbar}(k) = E_{j,\hbar}(k)\Psi_{j,\hbar}(k) \qquad (j \in \mathbb{N})$$

of its eigenvalue equations, the *band functions* $k \mapsto E_{j,\hbar}(k)$ are real analytic on $\mathbb{T}^*$, up to measure zero sets due to degeneracies. The *group velocity*

$$\overline{v}_{j,x_1}(k) := \hbar^{-1}\nabla E_{j,\hbar}(k)$$

maps Haar measure on $\mathbb{T}^*$ to some probability measure on $\mathbb{R}^d$. It is an easy exercise (see [AK1]) to show that $\overline{v}_{j,\hbar}$ determines the asymptotic velocity of the quantum particle in the periodic potential.

So the heuristic correspondence principle suggests that, after takting the mean $\nu_{E,\hbar}$ of these probability measures for j with $|E_{j,\hbar} - E| < \varepsilon(\hbar)$ (with $\lim_{\hbar \searrow 0} \varepsilon(\hbar) = 0$ at a suitable rate), we get weak convergence

$$\lim_{\hbar \searrow 0} \nu_{E,\hbar} = \nu_E \tag{6}$$

to the classical distribution of velocities.

However, fixing $k \in \mathbb{T}^*$, the corresponding statement is known to be *wrong*, due to tunneling in phase space. Up to now (6) has not been shown in full generality, but the following results have been obtained, partly in collaboration with J. Asch:

– (6) holds true if the classical motion is integrable or ergodic [Kn2, AK1]
– (6) holds true for the velocity distribution of the KAM tori, whose complement has only Liouville measure $\mathcal{O}(1/\sqrt{E})$ on the energy surface $\hat{H}^{-1}(E)$ if V is sufficiently smooth [AK2].

In the ergodic case,

– $\nu_E = \delta_0$, the Dirac measure at $0 \in \mathbb{R}^d$. So the particle is not ballistic with probability one.
– However, for $E > \sup_q V(q)$ the particle goes to infinity with probability one in the sense

$$\limsup_{t \to \infty} |q(t, x)| = \infty \quad \text{for} \quad \lambda_E\text{–almost every } x \in H^{-1}(E).$$

A concrete instance of such an ergodic behaviour is given by Coulombic $\mathcal{L}$–periodic potentials on $\mathbb{R}^2$, like

$$V(q) := \sum_{\ell \in \mathcal{L}} W(q - \ell)$$

with the Yukawa potential $W(q) := -\frac{e^{-\mu|q|}}{|q|}$.

Here for E larger than some threshold E_{th} we do not only have ergodicity but even *diffusion* on $\mathbb{R}^2$, that is the *diffusion constant*

$$D(E) := \lim_{T \to \infty} \frac{\langle (q(T, x) - q_0)^2 \rangle}{dT}$$

is finite and non-zero, if one takes the average $\langle \cdot \rangle$ with respect to a measure[12] of initial conditions $x \equiv (p_0, q_0)$ absolutely continuous w.r.t. Liouville measure λ_E on $H^{-1}(E)$, see Fig. 7. This was shown in [Kn1].

However, ergodicity does not preclude ballistic motion for a measure zero set of initial conditions. Indeed, as shown in [AK1], for $E > E_{\text{th}}$, and every

---

[12] $D(E)$ does not depend on the choice of that measure.

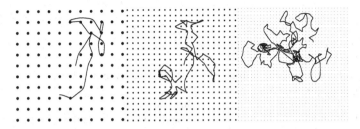

Fig. 7. Diffusive motion in a periodic potential with coulombic singularities

pair $\overline{w}^{\pm} \in \mathbb{R}^2$, $|\overline{w}^{\pm}| \leq \sqrt{E}$, there is an $x \in H^{-1}(E)$ with[13] $\overline{v}^-(x) = \overline{w}^-$ and $\overline{v}^+(x) = \overline{w}^+$.

It may be amusing that the dynamics realizes this four parameter set of asymptotic data on a measure zero set of initial conditions x in the three-dimensional manifold $H^{-1}(E)$.

In the above mentioned case, the motion on $\hat{H}^{-1}(E)$ is Anosov. What we would like to see is that a classical gas is ergodic, and this seems to be true in some sense for a gas of hard balls. However in the presence of smooth two-body interactions a similar result may be hard to prove, because of the following theorem:

Theorem 6 ([AK1]). *If $\hat{V} \in C^2(\mathbb{T}^d)$, then the hamiltonian motion generated by $\hat{H}(p,q) = \frac{1}{2}p^2 + \hat{V}(q)$ on $\hat{H}^{-1}(E)$ is never of Anosov type.*

References

[AM]   Abraham, R., Marsden, J.E.: Foundations of Mechanics. Reading: Benjamin 1978

[AK1]   J. Asch, A. Knauf: Motion in Periodic Potentials. Nonlinearity **11**, 175–200 (1998)

[AK2]   J. Asch, A. Knauf: Quantum Transport on KAM Tori. Commun. Math. Phys. **205**, 113–128 (1999)

[Ba]   W. Ballmann: Lectures on Spaces of Nonpositive Curvature. Birkhäuser 1995

[Bu]   L.A. Bunimovich et al: Dynamical Systems, Ergodic Theory and Applications. Encyclopaedia of Mathematical Sciences, Vol. 100. Springer 2000

[Fa]   K. Falconer: Techniques in Fractal Geometry. Wiley 1997

[KH]   A. Katok, B. Hasselblatt: Introduction to the Modern Theory of Dynamical Systems. Cambridge University Press 1997

[Ke]   G. Keller: Equilibrium States in Ergodic Theory. London Mathematical Society Student Texts 42. Cambridge University Press 1998

[Kl]   W. Klingenberg: Riemannian Geometry. De Gruyter 1995

---

[13] In fact this was shown only for a dense set, but by a limit argument holds true for all choices of $\overline{w}^+$ and $\overline{w}^-$.

[Kn1] A. Knauf: Ergodic and Topological Properties of Coulombic Periodic Potentials. Commun. Math. Phys. **110**, 89–112 (1987)

[Kn2] A. Knauf: Coulombic Periodic Potentials: The Quantum Case. Annals of Physics **191**, 205–240 (1989)

[LW]   C. Liverani, M.P. Wojtkowski: Ergodicity in Hamiltonian Systems. In: Dynamics Reported IV. Springer 1995

[Pa]   G. Paternain: Geodesic Flows. Birkhäuser 1999

[Wa]   P. Walters: An Introduction to Ergodic Theory. Graduate Texts in Mathematics, Vol. 79. Springer 1982

Number Theoretic Background

Zeév Rudnick

Raymond and Beverly Sackler School of Mathematical Sciences, Tel Aviv University, Tel Aviv 69978, Israel, *rudnick@post.tau.ac.il*

1 Introduction

This paper is an expanded version of some of the lectures given at the summer school in Bologna. In these lectures we gave an introduction to very basic number theory, assuming practically no background. The lectures were intended for graduate students in Math and Physics and while the material is completely standard, we tried to make the presentation as elementary as possible.

Some of the easier proofs are included, others are relegated to exercises, but several of the deeper facts are stated without proof. Most of the material may be found in classic texts such as [1] and [3].

2 Divisibility

2.1 Basics on Divisibility

We denote by $\mathbf{Z} = \{\dots, -2, -2, 0, 1, 2, \dots\}$ the set of integers.

The Euclidean property: If $b \neq 0$ then for any a, we can write

$$a = qb + r$$

with remainder $0 \le r < |b|$ and quotient q. [1]

Given a pair of integers a, b with $b \neq 0$, we say that b divides a, denoted as $b \mid a$, if the remainder $r = 0$, that is if $a = bq$ for some integer q. We will also say that b is a divisor of a.

The basic properties of the divisibility relation are

1. $b \mid 0$ for all nonzero b and $1 \mid a$ for all a.
2. Transitivity: $b \mid a$ and $c \mid b$ implies $c \mid a$.
3. if $d \mid a$ and $d \mid b$ then $d \mid ax + by$ for all integers x, y.
4. units: $a \mid 1$ if and only if $a = \pm 1$. Indeed, since nonzero integers have absolute value at least 1, the only solutions of the equation $xy = 1$ in integers are $(x, y) = (1, 1)$ or $(-1, -1)$.

---

[1] or $-|b|/2 < r \le |b|/2$.

5. For nonzero integers a, b we have both $a \mid b$ and $b \mid a$ if and only if $b = \pm a$. Indeed, if $b \mid a$ the we can write $a = bx$ for some integer x, and from $a \mid b$ we can write $b = ay$ for some $y \in \mathbf{Z}$. Thus $axy = a$ and since $a \neq 0$ this means $xy = 1$ which forces $x = \pm 1$.

2.2 The Greatest Common Divisor

Given a pair of integers a *common divisor* is an integer d which divides both. For instance the common divisors of 4 and 6 are the integers $\pm 1, \pm 2$.

Definition 1. *A greatest common divisor of a pair of nonzero integers a, b is a common divisor d which is maximal in the sense that if δ is any common divisor of a, b then $\delta \mid d$.*

An inspection shows that the greatest common divisors of 4 and 6 are ± 2.

The basic fact is the existence (not apriori obvious from our definition) and essential uniqueness of greatest common divisors. We will denote by $\gcd(a, b)$ a choice of greatest common divisor, which is unique if require that it be positive:

Theorem 1. *Any two nonzero integers a, b admit a greatest common divisor, unique up to sign. Furthermore, one can always find integers x, y so that $\gcd(a, b) = ax + by$.*

Below we will see a proof of this which gives an efficient algorithm for both computing the gcd and finding integer solutions of the equation $\gcd(a, b) = ax + by$.

We first derive a few properties of the gcd:

Definition 2. *A pair of integers a, b is coprime if $\gcd(a, b) = 1$*

A useful criterion for coprimality is

Lemma 1. *a, b are coprime if and only if there are integers x, y such that $ax + by = 1$.*

We will need to use the following:

Lemma 2. *a) If a, b are coprime and $a \mid bc$ then $a \mid c$.*
b) If a, b are coprime and both divide c then their product ab divides c.

Proof. Indeed, if a, b are coprime then we can write $1 = ax + by$ for integers x, y. Multiplying this equation by c we find

$$c = a \cdot xc + y \cdot bc . \tag{2.1}$$

For part (a), we are assuming $a \mid bc$ and so both summands on the RHS of (2.1) are divisible by a, hence so is the LHS, namely c.

For part (b), we are assuming that $a \mid c$ and so $ab \mid yb \cdot c$ and likewise since we assume that $b \mid c$, we have $ab \mid ax \cdot c$ and thus ab divides the LHS of (2.1) and so divides the LHS, namely c.

2.3 The Euclidean Algorithm

The Euclidean algorithm gives an efficient method for finding the gcd as well as for computing x, y so that $\gcd(a, b) = ax + by$. The method is as follows: Assume $|a| \geq |b|$. It will be convenient to set $r_{-1} := |a|$, and $r_0 := |b|$. We use division with remainder to write

$$a = q_1 b + r_1, \qquad 0 \leq r_1 < |b| \ .$$

If $r_1 = 0$ then $b \mid a$ and $\gcd(a, b) = b$. Otherwise, iterate this step with a replaced by b and b replaced by the remainder r_1 to write

$$b = q_2 r_1 + r_2, \quad 0 \leq r_2 < r_1 \ .$$

Continuing in this fashion, we get after k steps

$$r_{k-2} = q_k r_{k-1} + r_k, \qquad 0 \leq r_k < r_{k-1} \ .$$

Since the sequence of remainders $|b| = r_0 > r_1 > r_2 > \ldots$ is a strictly decreasing sequence of non-negative integers, this process has to terminate in a finite number of steps, say after n steps we have

$$r_{n-2} = q_n r_{n-1} + r_n, \qquad r_n \neq 0$$

and

$$r_{n-1} = q_{n+1} r_n$$

We claim that

$$\gcd(a, b) = r_n \ .$$

Moreover, the process gives integers x, y so that

$$\gcd(a, b) = ax + by \ .$$

To see this, we will show by descending induction on $i = n, n-1, \ldots, 0, -1$ that

$$r_i \mid r_n \tag{2.2}$$

and that

$$r_n = x_i r_{i-1} + y_i r_i \tag{2.3}$$

Once we have this, taking $i = 0, -1$ will give $r_n \mid r_0 = b$ and $r_n \mid r_{-1} = a$, and taking $i = 0$ in (2.3) will give $r_n = x_0 a + y_0 b$.

For $i = n$ we clearly have $r_n \mid r_n$ and $r_n = 1 r_n + 0 r_{n-1}$. For $i = n - 1$ we have $r_{n-1} = q_n r_n$ giving both (2.2) and (2.3) in that case. Assuming we know that $r_n \mid r_k$ and $r_n \mid r_{k-1}$, we use $r_{k-2} = q_k r_{k-1} + r_k$ to find that $r_n \mid r_{k-2}$. Further, assuming we know (2.3) for $i = k$ gives

$$\begin{aligned} r_n &= x_k r_{k-1} + y_k r_k \\ &= x_k r_{k-1} + y_k (r_{k-2} - q_k r_{k-1}) \\ &= y_k r_{k-2} + (x_k - q_k y_k) r_{k-1} \end{aligned}$$

that is (2.3) holds for $i = k - 1$ with $x_{k-1} = y_k$, $y_{k-1} = x_k - q_k y_k$.

Example: $a = 8$, $b = 5$: To compute $\gcd(8, 5)$, we carry out the following steps:

$$8 = 1 \cdot 5 + 3$$
$$5 = 1 \cdot 3 + 2$$
$$3 = 1 \cdot 2 + 1$$
$$2 = 2 \cdot 1 \,.$$

Thus $\gcd(8, 5) = 1$. To find integer solutions of $8x + 5y = 1$, we proceed backwards:

$$1 = 3 - 1 \cdot 2$$
$$= 3 - (5 - 1 \cdot 3) = 2 \cdot 3 - 5$$
$$= 2 \cdot (8 - 1 \cdot 5) - 5 = 2 \cdot 8 - 3 \cdot 5$$

and thus we found the solution $x = 2$, $y = -3$ to $8x + 5y = 1$.

An examination of the Euclidean algorithm shows that the number of steps is at most $2 \log_2 |b| + 1 = O(\log \min(|a|, |b|))$, that is the complexity is **linear** in the number of bits needed to represent the input.

Exercise 1. Prove this estimate on the number of steps.

Hint: Show that $r_k \leq r_{k-2}/2$.

Exercise 2. Show that one cannot significantly improve this estimate.

Hint: Take a, b to be consecutive Fibonacci numbers. These are defined recursively as $F_0 = 0$, $F_1 = 1$, and for $n \geq 1$ by $F_{n+1} = F_n + F_{n-1}$. Show that the number of steps for computing $\gcd(F_{n+1}, F_n)$ is n (?) and that $n \sim \log F_n / \log((1 + \sqrt{5})/2)$ as $n \to \infty$.

2.4 The Diophantine Equation $ax + by = c$

Given integers a, b, c, we wish to find *integer* solutions to the equation

$$ax + by = c \,.$$

As the example $4x + 6y = 1$ illustrates, such solutions need not exist, since the RHS is odd while the LHS is even! More generally, if $\gcd(a, b)$ does not divide c then there will be no integer solutions. It turns out that this divisibility condition is the only obstruction to the existence of solutions, and once this obstruction vanishes then there are infinitely many integer solutions:

Theorem 2. *The equation $ax + by = c$ has integer solutions if and only if $\gcd(a, b) \mid c$.*

If there is one solution (x_0, y_0) then there are infinitely many integer solutions, and they are all given by

$$x_k = x_0 + k\frac{b}{\gcd(a,b)}, \quad y_k = y_0 - k\frac{a}{\gcd(a,b)}$$

where k runs over all integers.

The proof of this theorem is very easy once we know the existence of one solution. To find a solution, first use the Euclidean algorithm to solve the equation $au + bv = \gcd(a, b)$ and then take

$$x_0 = u\frac{c}{\gcd(a,b)}, \quad y_0 = v\frac{c}{\gcd(a,b)} \ .$$

3 Prime Numbers

3.1 The Fundamental Theorem of Arithmetic

A *prime* is a natural number $p > 1$ which has no proper divisors (that is no divisors other than $\pm p$ and ± 1).

The sequence of primes is thus $p_1 = 2$, $p_2 = 3$, $p_3 = 5, \ldots$.

Lemma 3. *If p is prime which divides a product: $p \mid bc$, then it has to divide one of the factors: $p \mid b$ or $p \mid c$.*

Indeed, if p is prime and p does not divide b then automatically p and b are coprime, hence the result follows from Lemma 2.

We claim that every integer $n > 1$ is a product of primes: Indeed, $n > 1$ is either a prime, in which case we are done, or decomposable: $n = ab$, with $a, b > 1$. In the latter case, we have $a, b < n$ and arguing by induction, both a, b are products of primes and hence so is $n = ab$.

Since every integer is a product of primes, they are thus the building blocks of all the integers. In fact, more is true - the factorization into products of primes is *unique*:

Theorem 3 (Fundamental Theorem of Arithmetic). *Every natural number is* uniquely *decomposable into a product of prime powers.*

Exercise 3. Show that if the prime factorization of a pair of integers is given by $a = \prod p^{\alpha(p)}$ and $b = \prod p^{\beta(p)}$ then

$$\gcd(a,b) = \prod p^{\min(\alpha(p), \beta(p))} \ .$$

While we know that every integer factors into a product of primes, a basic problem is how to find this factorization efficiently. Currently, there is no known algorithm which will give the prime factorization of an integer in a number of steps which is polynomial in the input (quantum algorithms aside).

3.2 There Are Infinitely Many Primes

Theorem 4 (Euclid). *There are infinitely many primes.*

Proof. Argue by *reductio ad absurdum*: If there were finitely many primes, say only M of them, then form the integer $Q = p_1 \cdot p_2 \cdots \cdots p_M + 1$. It is either a prime or decomposable. Since Q is greater than all the primes $p_1, \ldots, p_M$, it cannot be a prime. However, Q clearly leaves remainder 1 on division by each of the available primes p_i, and thus being divisible by no prime, cannot decompose into a product of primes! We thus arrive at a contradiction.

Exercise 4. Show that there infinitely many primes of the form $4k + 3$.

3.3 The Density of Primes

After knowledge that there are infinitely many primes, one can try to assess their density. Gauss recounted that in 1792, as a boy of 15, he arrived at the conjecture that the density of primes near x is about $1/\log x$ and so if we denote by $\pi(x)$ the number of primes up to x

$$\pi(x) := \#\{n : p_n \leq x\}$$

then $\pi(x)$ is asymptotically equal to the logarithmic integral, given for $x > 2$ by

$$\mathrm{Li}(x) := \int_2^x \frac{dt}{\log t}$$

In turn, $\mathrm{Li}(x)$ has an asymptotic expansion

$$\mathrm{Li}(x) = \frac{x}{\log x} + \frac{x}{(\log x)^2} + \cdots + c_n \frac{x}{(\log x)^n} + O\left(\frac{x}{(\log x)^{n+1}}\right)$$

To check the strength of $\mathrm{Li}(x)$ as an approximation to $\pi(x)$, we examine Table 1 (writing $[y] :=$ integer part of y). As is seen from this table, $\mathrm{Li}(x)$ is a remarkably good approximation to $\pi(x)$ in this range. As a measure of the quality of the approximation, note that the width of the third column is

Table 1. Comparison between $\pi(x)$ and $\mathrm{Li}(x)$.

x	$\pi(x)$	$[\mathrm{Li}(x)] - \pi(x)$
10^8	5,761,455	754
10^{10}	455,052,511	3,104
10^{12}	37,607,912,018	38,263
10^{14}	3,204,941,750,802	314,890
10^{16}	279,238,341,033,925	3,214,632

about a half of the width of the second one, that is to say that the remainder is approximately square root of the main term!

The statement that $\pi(x) \sim \text{Li}(x)$ is known as the Prime Number Theorem. It was proved in 1896 by Hadamard and de la Vallée Poussin, by using the Riemann zeta function. The empirical statement made above from the data in Table 1 as to the magnitude of the remainder in this approximation is a form of the celebrated Riemann Hypothesis, see Sect. 8.2.

3.4 Primes in Arithmetic Progressions

An important issue is the existence of primes in a given arithmetic progression: Given a and $q > 1$, to find a large prime p with $p \equiv a \mod q$. Clearly, in some instances it cannot be done, say the progression $\{2, 4, 6, 8, \dots\}$ contains no large primes as all primes except 2 are odd. Likewise, if a and q have a common factor $d > 1$ then it divides every element of the progression $a, a + q, a + 2q \dots$ and so there are no primes in it (excepting perhaps if $a = d$ is prime). We should thus restrict attention to the case that a and q are co-prime. It turns out that this is the only obstruction to the existence of primes in arithmetic progression, as was proved by Dirichlet in 1837. In fact there are arbitrarily large primes in every progression not excluded by such reasoning:

Theorem 5 (Dirichlet's Theorem). *For $q > 1$ and any a co-prime to q, there are infinitely many primes of the form $a + kq$.*

One can try to give an argument for this along the lines of Euclid's argument for the existence of infinitely many primes (Theorem 4). This works in a few cases of small q, and for some special progressions such as $p \equiv 1 \mod q$, but this line of attack has not yielded Dirichlet's theorem in its full force.

A quantitative version of Dirichlet's theorem is the Prime Number Theorem for arithmetic progressions, which asserts that for fixed $q > 1$, every progression $a \mod q$ has asymptotically the same density of primes. Thus setting $\pi(x; q, a) := \#\{p \leq x : p \equiv a \mod q\}$, we have for a coprime to q

$$\pi(x; q, a) \sim \frac{1}{\phi(q)} \text{Li}(x) .$$

4 Continued Fractions

Continued fractions are expressions such as

$$1 + \cfrac{1}{2 + \cfrac{1}{3 + \cfrac{1}{4}}} .$$

We now study them systematically.

Given integers $a_0 \in \mathbf{Z}$, $a_1, a_2, \cdots \geq 1$, consider the finite continued fraction

$$[a_0; a_1, \ldots, a_m] := a_0 + \cfrac{1}{a_1 + \cfrac{1}{a_2 + \cfrac{1}{\ddots + \cfrac{1}{a_m}}}}.$$

To compute this fraction, one defines integers p_m, q_m by the recursion ($m \geq 1$):

$$p_m = a_m p_{m-1} + p_{m-2}$$
$$q_m = a_m q_{m-1} + q_{m-2}$$

with $p_{-1} = 1$, $p_0 = a_0$, $q_{-1} = 0$, $q_0 = 1$. These satisfy the relations

$$p_m q_{m-1} - p_{m-1} q_m = (-1)^{m-1}$$

and

$$p_m q_{m-2} - p_{m-2} q_m = (-1)^m a_m \; .$$

On then shows that

$$[a_0; a_1, \ldots, a_m] = p_m/q_m \; .$$

The infinite simple continued fraction $[a_0; a_1, a_2, \ldots]$ is the limit of the "convergents" p_m/q_m. Every irrational α has a unique continued fraction expansion.

Example: We will obtain the continued fraction expansion of the quadratic irrationality $\sqrt{3}$: Since $\sqrt{3}$ lies between 1 and 2, we write $\sqrt{3} = 1 + 1/x_1$ with

$$x_1 = \frac{1}{\sqrt{3} - 1} = \frac{\sqrt{3} + 1}{2}$$

Thus $x_1 = 1 + 1/x_2$ with

$$x_2 = \frac{2}{\sqrt{3} - 1} = \sqrt{3} + 1$$

and thus we can write $x_2 = 2 + 1/x_3$ with

$$x_3 = \frac{1}{\sqrt{3} - 1} = x_1 \; .$$

The procedure has thus cycled back and we may continue it indefinitely to find that $\sqrt{3}$ has the *periodic* continued fraction expansion

$$\sqrt{3} = 1 + \cfrac{1}{1 + \cfrac{1}{2 + \cfrac{1}{\ddots}}} = [1; 1, 2, 1, 2, \ldots] = [1; \overline{1, 2}] \ .$$

This is something of a rarity; it turns out that an irrational has a periodic continued fraction expansion if and only if it is a *quadratic irrationality*, that is of the form $r + s\sqrt{d}$, with r, s rational and $d > 1$ an integer which is not a perfect square.

The convergents give very good rational approximations to α: We have

$$\frac{1}{2} \frac{1}{q_m q_{m+1}} < |\alpha - \frac{p_m}{q_m}| < \frac{1}{q_m q_{m+1}} \ .$$

The convergents p_m/q_m are the "best" rational approximations to α, in the following senses: If p/q satisfies $|\alpha - p/q| < 1/2q^2$ then $p/q = p_m/q_m$ for some m. Moreover, for $m > 1$, if $0 < q \leq q_m$ and $p/q \neq p_m/q_m$ then $|\alpha - p/q| > |\alpha - p_m/q_m|$.

5 Modular Arithmetic

5.1 Congruences

Given $N > 1$, we say that two integers a, b are congruent modulo N, written as $a \equiv b \mod N$, if $N \mid a - b$.

Congruence modulo N is an "equivalence relation" on the set of integers, that is as follows immediately from the definition,

1. $a \equiv a \mod N$
2. $a \equiv b \mod N$ if and only if $b \equiv a \mod N$
3. $a \equiv b \mod N$ and $b \equiv c \mod N$ implies $a \equiv c \mod N$

Given $N > 1$, every integer is congruent to precisely one of the N integers $\{0, 1, \ldots, N - 1\}$, as is seen by writing $a = qN + r$ with remainder $0 \leq r < N$. Thus denoting by $\mathbf{Z}/N\mathbf{Z}$ the set of congruence classes of integers modulo N, we see that a complete set of representatives can be taken to be $\{0, 1, \ldots, N - 1\}$, though other choices are equally valid.

The set of congruence classes modulo N also admits algebraic operations of addition and multiplication. To see this, one needs to check that if $a \equiv a'$ mod N and $b \equiv b' \mod N$ then $a+b \equiv a'+b' \mod N$ and likewise $a \cdot b \equiv a' \cdot b'$ mod N, that is the sum/product of congruence classes is independent of the choice of representatives.

Example: See Tables 2 and 3 for the multiplication tables modulo 4 and 5.

The existence of addition and multiplication satisfy the usual laws of integer arithmetic, that is commutativity, associativity, distributivity etc. This is formally expressed by saying that $\mathbf{Z}/N\mathbf{Z}$ is a *"ring"*.

Table 2. Multiplication modulo 4

	0	1	2	3
0	0	0	0	0
1	0	1	2	3
2	0	2	0	2
3	0	3	2	1

Table 3. Multiplication modulo 5

	0	1	2	3	4
0	0	0	0	0	0
1	0	1	2	3	4
2	0	2	4	1	3
3	0	3	1	4	2
4	0	4	3	2	1

5.2 Modular Inverses

An integer a is *invertible modulo* N if there is an integer b so that $ab \equiv 1$ mod N. We say that b is an inverse of a modulo N, denoted by $b \equiv a^{-1}$ mod N.

For instance 1 is always invertible and $1^{-1} \equiv 1$ mod N. Invertibility makes sense for congruence classes modulo N (why?). We will denote the set of invertible residue classes modulo N by $(\mathbf{Z}/N\mathbf{Z})^*$.

Examining the multiplication tables (2), (3), we see that modulo 4, the invertible residue classes are $1, 3$ with $3^{-1} \equiv 3$ mod 4 while all nonzero residue classes modulo 5 are invertible, and $2^{-1} \equiv 3$ mod 5, $3^{-1} \equiv 2$ mod 5 and $4^{-1} \equiv 4$ mod 5.

As these examples indicate, an inverse modulo N, of it exists, is *unique* modulo N. Indeed, if $ab \equiv 1$ mod N and $ac \equiv 1$ mod N then using commutativity and associativity of modular multiplication we have

$$b \equiv b \cdot 1 \equiv b \cdot ac \equiv (ab) \cdot c \equiv 1 \cdot c \equiv c \mod N$$

and thus $b \equiv c$ mod N as claimed.

Moreover the product of invertible classes is still invertible. Thus we get a commutative group structure on the set $(\mathbf{Z}/N\mathbf{Z})^*$ of invertible residue classes modulo N. We will hence refer to $(\mathbf{Z}/N\mathbf{Z})^*$ as the *multiplicative group modulo* N. save

A criterion for invertibility modulo N, as well as an algorithm for finding the modular inverse, is given by

Lemma 4. *A necessary and sufficient condition for an integer a to be invertible modulo N is that a and N are coprime:* $\gcd(a, N) = 1$.

Indeed, a and N are coprime if and only if we can solve $ax + Ny = 1$ (Lemma 1), and the latter equation is equivalent to solubility of the congruence $ax \equiv 1 \mod N$.

As an immediate corollary, we see that if N is *prime* then all nonzero residue classes modulo N are invertible, since the only divisors of N are 1 and N and so the only alternative to $\gcd(a, N) = 1$ in this case is $\gcd(a, N) = N$, and thus $a \equiv 0 \mod N$.

Moreover, as described in (2.3), the Euclidean algorithm provides for an efficient method of finding a solution of $ax + Ny = 1$, that is of polynomial time in the input, and thus an efficient method for finding modular inverses.

Definition 3. *Euler's totient function $\phi(N)$ is the number of invertible residue classes modulo N.*

As we saw above, of p is prime then all nonzero residue classes mod p are invertible and thus $\phi(p) = p - 1$ in this case.

Exercise 5. Show that for prime p we have $\phi(p^k) = p^k - p^{k-1}$ $(k \geq 1)$.

5.3 The Chinese Remainder Theorem

Given $x \mod mn$, we get a pair of residue classes $(x \mod m, x \mod n)$ in $\mathbf{Z}/m\mathbf{Z} \times \mathbf{Z}/n\mathbf{Z}$. If m, n are coprime, then we may recover $x \mod mn$ from this pair.

Theorem 6 (CRT). *If m, n are coprime then for a, b we can solve the congruence*

$$\begin{cases} x \equiv a \mod m \\ x \equiv b \mod n \end{cases} \tag{5.1}$$

the solution is unique modulo mn.

Proof. To see uniqueness, note that if x, x' are solutions of (5.1), then both m and n divide $x - x'$ and since m, n are coprime this forces $mn \mid x - x'$, that is $x \equiv x' \mod mn$.

The method of solution is constructive: Since m, n are coprime, m is invertible modulo n. Denote by $\bar{m}$ the inverse of $m \mod n$, and likewise let $\bar{n}$ be the inverse of $n \mod m$. Then a solution of the system of congruences (5.1) is given by

$$x = n\bar{n}a + m\bar{m}b \mod mn .$$

Indeed, modulo m we have $n\bar{n} \equiv 1 \mod m$ while $m \equiv 0 \mod m$ and thus $x = n\bar{n}a + m\bar{m}b \equiv 1 \cdot a + 0 \cdot \bar{m}b \equiv a \mod m$, and likewise $x \equiv b \mod n$.

Exercise 6. If $x \equiv a \mod 3$, $x \equiv b \mod 5$, $x \equiv c \mod 7$ find $x \mod 105$.

Exercise 7. Show that if m, n are coprime then x is invertible modulo mn if and only if it is invertible both modulo m and modulo n.

Exercise 8. Show that if m, n are coprime then $\phi(mn) = \phi(m)\phi(n)$.

Exercise 9. Show that for $N > 1$ we have $\phi(N) = N \prod_{p|N}(1 - 1/p)$ where the product is over all primes dividing N.

5.4 The Structure of the Multiplicative Group $(\mathbf{Z}/N\mathbf{Z})^*$

In this section, we will investigate the structure of multiplication in the group of invertible residue classes modulo N.

Exercise 10. If $a, b \in \mathbf{Z}/N\mathbf{Z}$ then we may compute $a \cdot b \mod N$ in $O(\log a \log b)$ bit operations.

A fundamental aspect is the power operation, of raising an element to a power: $a \mapsto a^k \mod N$. It turns out that this is computationally easy:

Exercise 11 (Divide and conquer). Show that we can compute $a^k \mod N$ in at most $O(\log k (\log N)^2)$ elementary bit operations.

Definition 4. *The order of $a \in (\mathbf{Z}/N\mathbf{Z})^*$ is the least integer $k \geq 1$ for which $a^k \equiv 1 \mod N$.*

We denote this integer by $\mathrm{ord}(a, N)$.

Exercise 12. Make up tables of orders of all elements modulo $5, 8, 11$.

That $\mathrm{ord}(a, N)$ exists is guaranteed by:

Theorem 7 (Fermat-Euler). *For any $a \in (\mathbf{Z}/N\mathbf{Z})^*$ we have*

$$a^{\phi(N)} \equiv 1 \mod N .$$

In particular, we see that $\mathrm{ord}(a, N) \leq \phi(N)$. In fact more is true:

Proposition 1. *Let $a \in (\mathbf{Z}/N\mathbf{Z})^*$ be invertible modulo N.*
  a) Suppose $a^k \equiv 1 \mod N$. Then $\mathrm{ord}(a, N)$ divides k.
  b) In particular, $\mathrm{ord}(a, N)$ divides $\phi(N)$.

Proof. Write $k = q\,\mathrm{ord}(a, N) + r$, with $0 \leq r < \mathrm{ord}(a, N)$. Then $1 = a^k = (a^{\mathrm{ord}(a,N)})^q \cdot a^r \equiv 1^q \cdot a^r = a^r \mod N$. Since $r < \mathrm{ord}(a, N)$, this forces $r = 0$ by the minimality of $\mathrm{ord}(a, N)$, that is $\mathrm{ord}(a, N) \mid k$. Part (b) follows from part (a) and Fermat-Euler.

Exercise 13. Let $p \neq 2, 5$ be a prime, and consider the decimal (base 10) expansion of the rational $1/p$. This expansion is periodic of the form $1/p = 0.\overline{a_1 \dots a_T}$ where T denotes the minimal period. For instance, $1/3 = 0.333\dots = 0.\overline{3}$ $(T = 1)$, $1/7 = 0.\overline{142857}$ $(T = 6)$, $1/11 = 0.\overline{09}$ $(T = 2)$ etc.
  Show that $T = \mathrm{ord}(10, p)$. Generalize.

5.5 Primitive Roots

The maximal order of an invertible element is $\phi(N)$. We will say that $a \in (\mathbf{Z}/N\mathbf{Z})^*$ is a *primitive root modulo N* if $\mathrm{ord}(a, N) = \phi(N)$.

An examination of the tables of orders of elements modulo 5, 8 ,... (see Tables 4, 5 and 6) indicates that this sometimes does indeed happen, though when the modulus is 8 the maximal order is 2 rather than $4 = \phi(8)$.

The following theorem explains this empirical finding:

Theorem 8. *If p is a prime then there is a primitive root modulo p.*

For composite moduli, it is relatively rare to have primitive roots:

Exercise 14. a) Show that if $n > 2$ then $\phi(n)$ is even.

b) If $m, n > 2$ are coprime then there is no primitive root modulo mn.

It turns out that there is a primitive root modulo N if and only if $N = 2, 4$ or $N = p^k, 2p^k$ where p is an odd prime.

While Theorem 8 guarantees the existence of a primitive root modulo a prime, one does not know of an efficient algorithm that given a (large) prime p, finds a primitive root modulo p.

In this context, there is a conjecture of Emil Artin from the 1920's which among other things says that the number 2 is a primitive root modulo infinitely many primes, and the same is true for any integer $a \neq \pm$ and not a perfect square - see the survey by Ram Murty [4] for further details.

The importance of the notion of primitive roots comes from the following observation: For $g \in (\mathbf{Z}/N\mathbf{Z})^*$, the function $x \mapsto g^x \bmod N$ has period exactly $\mathrm{ord}(g, N)$ by Proposition 1, and in particular if g is a primitive root modulo N then this period is precisely $\phi(N)$. Thus if g is a primitive root modulo N then for any $a \in (\mathbf{Z}/N\mathbf{Z})^*$ we may write $a \equiv g^x \bmod N$ with x unique modulo $\phi(N)$. Since there are $\phi(N)$ invertible residues and the same number of (necessarily invertible) powers $g^x \bmod N$, we find:

Table 4. Orders of elements mod 5

a	1 2 3 4
$\mathrm{ord}(a, 5)$	1 4 4 2

Table 5. Orders of elements mod 8

a	1 3 5 7
$\mathrm{ord}(a, 8)$	1 2 2 2

Table 6. Orders of elements mod 11

a	1 2 3 4 5 6 7 8 9 10
$\mathrm{ord}(a, 11)$	1 10 5 5 5 10 10 10 5 2

Lemma 5. *An invertible element* $g \in (\mathbf{Z}/N\mathbf{Z})^*$ *is a primitive root modulo* N *if and only if every element* $a \in (\mathbf{Z}/N\mathbf{Z})^*$ *can be written as* $a \equiv g^x \mod N$.

This lemma allows us to convert modular *multiplication* in the group $(\mathbf{Z}/N\mathbf{Z})^*$ into modular *addition* in $\mathbf{Z}/\phi(N)\mathbf{Z}$, assuming we have a primitive root, since if $a \equiv g^x$ and $b \equiv g^y$ then $a \cdot b \equiv g^{x+y} \mod N$.

If $a \equiv g^x \mod N$, we will write $x := \mathrm{Ind}_g(a, N)$. We may think of $\mathrm{Ind}_g(a, N)$ as a *discrete logarithm*, since by the above reasoning $\mathrm{Ind}_g(ab) \equiv \mathrm{Ind}_g(a) + \mathrm{Ind}_g(b) \mod \phi(N)$. Given g, x it is easy to compute $g^x \mod N$, there is no known efficient method of determining $x = \mathrm{Ind}_g(a, N)$ from a, g and N. This is known as the *discrete logarithm problem*.

6 Quadratic Congruences

6.1 Euler's Criterion

Let p be an odd prime. We will study the congruence

$$x^2 \equiv a \mod p \tag{6.1}$$

where a is invertible modulo p. One algorithm for deciding the solubility of this congruence is given by Euler's criterion:

Proposition 2 (Euler's Criterion). *There is a solution of* $x^2 \equiv a \mod p$ *if and only if* $a^{(p-1)/2} \equiv 1 \mod p$.

Proof. Indeed, if $a \equiv x^2 \mod p$ then $a^{(p-1)/2} = x^{p-1} \equiv 1 \mod p$. For the reverse direction, we may use the existence of a primitive root $g \mod p$ to try and solve the congruence (6.1) by writing

$$a \equiv g^b \mod p, \qquad 0 \le b < p - 1$$

It will suffice to show that $b \equiv 2z \mod p-1$ since then $x \equiv g^z \mod p$ solves the congruence (6.1). Substituting $a \equiv g^b \mod p$ in $a^{(p-1)/2} \equiv 1 \mod p$ gives

$$g^{\frac{p-1}{2}b} = 1 \mod p.$$

By Proposition 1 this forces $(p-1)b/2 = 0 \mod \mathrm{ord}(g, p)$ and since g is a primitive root, $\mathrm{ord}(g, p) = p-1$ and we find that $b \equiv 0 \mod 2$, that is $b = 2z$ as required.

As an immediate consequence, we see that for p odd, $-1 \equiv x^2 \mod p$ if and only if $p \equiv 1 \mod 4$, since $(-1)^{(p-1)/2} = 1$ precisely in that case.

6.2 The Legendre Symbol and Quadratic Reciprocity

For $p \neq 2$ an odd prime and a invertible modulo p, the Legendre symbol $\left(\frac{a}{p}\right)$ is defined as

$$\left(\frac{a}{p}\right) = \begin{cases} +1 & a \equiv x^2 \mod p \\ -1 & \text{otherwise} \end{cases}$$

It is sometime convenient to extend the definition to include non-invertible residues by requiring $\left(\frac{a}{p}\right) = 0$ if $p \mid a$.

Below are some simple properties of the Legendre symbol:

1. If $a \equiv b \mod p$ then $\left(\frac{a}{p}\right) = \left(\frac{b}{p}\right)$.

2. $\left(\frac{x^2}{p}\right) = +1$. Both these follow from the very definition of the Legendre symbol.

3. Euler's criterion can be reformulated to read

$$\left(\frac{a}{p}\right) \equiv a^{(p-1)/2} \mod p$$

This gives a computationally efficient method of finding the Legendre symbol, as we can compute $a^{(p-1)/2} \mod p$ in $O(\log^3 p)$ steps.

A simple consequence is a rule for when -1 is a square modulo p:

$$\left(\frac{-1}{p}\right) = (-1)^{(p-1)/2} = \begin{cases} +1, & p \equiv 1 \mod 4 \\ -1, & p \equiv 3 \mod 4 \end{cases}$$

4. Multiplicativity:

$$\left(\frac{ab}{p}\right) = \left(\frac{a}{p}\right)\left(\frac{b}{p}\right)$$

This follows from Euler's criterion!

More profound is the celebrated law of Quadratic Reciprocity, conjectured by Euler and proved by Gauss:

Theorem 9. *If $p \neq q$ are distinct odd primes then*

$$\left(\frac{p}{q}\right)\left(\frac{q}{p}\right) = (-1)^{\frac{p-1}{2}\cdot\frac{q-1}{2}}$$

An addendum is the quadratic character of 2:

$$\left(\frac{2}{p}\right) = (-1)^{(p^2-1)/8} = \begin{cases} +1, & p \equiv \pm 1 \mod 8 \\ -1, & p \equiv \pm 3 \mod 8 \end{cases}$$

Example: To illustrate the power of the law of quadratic reciprocity, we will compute the Legendre symbol $\left(\frac{5}{p}\right)$ for primes $p \neq 2, 5$. We have

$$\left(\frac{5}{p}\right) = \left(\frac{p}{5}\right)(-1)^{(p-1)/2\cdot(5-1)/2} = \left(\frac{p}{5}\right).$$

To compute $\left(\frac{p}{5}\right)$, we only need to know the remainder of p after division by 5 and then check that the invertible squares modulo 5 are 1 and 4. This gives that $\left(\frac{5}{p}\right) = 1$ if $p \equiv \pm1 \mod 5$ and $\left(\frac{5}{p}\right) = -1$ for $p \equiv \pm2 \mod 5$.

7 Pell's Equation

Given a positive integer d, not a perfect square, we wish to find the integer solutions of the equation

$$x^2 - dy^2 = 1 .$$

This equation was studied several centuries ago. Its study in modern times was championed by Fermat (Pell had nothing to do with this equation, and owes it being named after him to Euler).

Exercise 15. Show that if d is a perfect square then the equation $x^- dy^2 = 1$ has only finitely many integer solutions.

Obvious solutions are $(x,y) = (\pm1,0)$, called the *trivial solutions*. The nontrivial solutions come in quadruples: If (x,y) are solutions then so are $(\pm x, \pm y)$. We will say that a solution is *positive* if $x, y > 0$.

It turns out that nontrivial solutions always exist, though they are quite sparse. We may clearly order the positive solutions by increasing size of their y coordinate, or equivalently by the size of their x-coordinate. The first non-trivial solution (x_1, y_1) will be called the *fundamental solution*.

Example: Suppose $d = 2$. To find (positive) solutions of the corresponding Pell equation, we rewrite it as

$$1 + dy^2 = x^2$$

and the proceed to search through values of $y = 1, 2, \ldots$, to find those for which $1 + dy^2$ is a perfect square. This process quickly yields the solutions $(x,y) = (3,2), (17,12), (99,70), (577,408), \ldots$. In this way we can clearly find all solutions up to any given value of y.

In Table 7 we enumerate the y-coordinate of first few positive solutions (x_n, y_n) in the case $d = 2$. One can clearly see an exponential increase of y_n with n.

Here it was easy to find the first few solutions by a brute-force search. This is not always the case. For instance, for $d = 61$ the equation was already treated in the 12-th century in India, and it was found that the fundamental solution has $y_1 = 226,153,980$. It is unlikely that this was found by hand merely by brute force!

Another example of a large fundamental solution for a small value of d is given by $d = 109$, when $y_1 = 15,140,424,455,100$.

Table 7. Solutions of $x^2 - 2y^2 = 1$

n	y_n
1	2
2	12
3	70
4	408
5	2378
6	13860
7	80782
8	470832
9	2744210
10	15994428

7.1 The Group Law

A remarkable feature of the Pell equation is the existence of a composition law on the set of solutions, turning them into a commutative group. One way of seeing this is to first endow the set of *real* solutions with a group law. Geometrically, the real solutions form a hyperbola $x^2 - dy^2 = 1$ with two sheets, and we do this just for the right sheet $(x > 0)$, via the following parameterization by means of the hyperbolic functions:

$$x(t) := \cosh(t), \quad y(t) := \frac{\sinh(t)}{\sqrt{d}}$$

If we denote by $P_t = (x(t), y(t))$ the point corresponding to t then the group law is the one inherited from the additive group of the reals, namely

$$P(t) * P(t') := P(t + t') .$$

More generally, we can write any real solution as $\pm P(t)$ and then declare the group law to be $\epsilon P(t) * \epsilon' P(t') := \epsilon\epsilon' P(t + t')$, $\epsilon, \epsilon' = \pm 1$.

In this form this does little except to demonstrate the existence of the group law, since recovering $t + t'$ from t and t' involves a transcendental inversion problem. However, we may use the addition formulae for the hyperbolic functions to compute the x and y coordinates of the composition. If we set $P = (x, y)$, $P' = (x', y')$ then $P * P' = (x'', y'')$ with

$$x'' = xx' + dyy', \qquad y'' = xy' + x'y . \tag{7.1}$$

In particular, the inverse of (x, y) is $(x, -y)$.

Note that from (7.1) it is not transparent to see that the addition law is associative!

In the form (7.1) we immediately see that the composition of *rational* solutions is still rational and ditto for *integer* solutions, which are our goal. Thus we see that the integer solutions form a group.

An easy way to recall the composition law (7.1) for rational solutions is to map them to quadratic irrationalities: $(x, y) \mapsto \alpha = x + \sqrt{d}y$, in which case composition is given by ordinary multiplication.

Rational solutions are easy to find by the "secant method".

Exercise 16. Show that all rational solutions of $x^2 - dy^2 = 1$ are given by $(-1, 0)$ together with

$$\{(\frac{1 + dt^2}{1 - dt^2}, \frac{2t}{1 - dt^2}) : t \text{ rational}\} . \tag{7.2}$$

Hint: Start with the trivial solution $(-1, 0)$. Given a point $(x, y) \neq (-1, 0)$ on the hyperbola $x^2 - dy^2 = 1$, draw the line connecting the two points $(-1, 0)$ and (x, y). This line will intersect the y-axis at a point $(0, t)$. Show that $y = t(x + 1)$ and substitute back into the original equation $x^2 - dy^2 = 1$ to find the expression (7.2). Argue that as t varies over all rationals, we get all the rational solutions other than $(-1, 0)$.

7.2 Integer Solutions

As we saw on the basis of examples, integer solutions are harder to come by. From the shape of the composition law (7.1) we see that if there is one nontrivial integer solution, then there are automatically *infinitely many* integer solutions: We may assume that we have a positive solution $P = (x_1, y_1)$, $x_1, y_1 > 0$ and then composing it with itself we get the solutions $P^{*2} = P * P$, ..., $P^{*n} = P^{*(n-1)} * P = (x_n, y_n)$ and from (7.1) its is clear that $x_n \to \infty$ as $n \to \infty$.

Theorem 10. *a) If $d > 0$ is not a perfect square then there are infinitely many integer solutions of Pell's equation $x^2 - dy^2 = 1$.*

b) If we denote by $\epsilon_d = (x_1, y_1)$ the fundamental solution then all integer solutions are given by $x + \sqrt{d}y = \pm\epsilon_d^n$, $n \in \mathbf{Z}$.

7.3 Finding the Fundamental Solution

While Theorem 10 guarantees the existence of solutions and that all solutions are found from knowledge of the fundamental solution, it tells us nothing about how to find the fundamental solution. One method of course is to search as described above. However there is an alternative method which is more efficient and involves the continued fraction expansion of $\sqrt{d}$. The situation is as follows: The continued fraction expansion of $\sqrt{d}$ is periodic, of the form:

$$\sqrt{d} = [a_0; \overline{a_1, \ldots, a_h}]$$

where h is the minimal period of the expansion. Let p_{h-1}/q_{h-1} be the $(h-1)$-st partial convergent. Then

$$p_{h-1}^2 - dq_{h-1}^2 = (-1)^h.$$

If h is *even* then the fundamental solution is (p_{h-1}, q_{h-1}).

If h is *odd* then the fundamental solution ϵ_d is given by the $(2h-1)$-st partial convergent, and moreover

$$\epsilon_d = p_{2h-1} + \sqrt{d}q_{2h-1} = (p_{h-1} + \sqrt{d}q_{h-1})^2.$$

Examples: $d = 7$: Then $\sqrt{7} = [2; \overline{1,1,1,4}]$ has period $h = 4$. The 3rd partial convergent is $[2; \overline{1,1,1}] = 8/3$ and the fundamental solution is $(8, 3)$.

$d = 61$: Then $\sqrt{61} = [7; \overline{1,4,3,1,2,2,1,3,4,1,14}]$ and this allows us to compute the (large) fundamental solution.

8 The Riemann Zeta Function

The Riemann zeta function is defined for complex s with $\operatorname{Re}(s) > 1$ by the series

$$\zeta(s) = \sum_{n=1}^{\infty} \frac{1}{n^s}.$$

We give an introduction to its basic properties (see [2]).

A basic fact is Euler's product formula, which displays the connection between $\zeta(s)$ and primes:

Theorem 11. *For $\operatorname{Re}(s) > 1$, $\zeta(s)$ can be represented by the convergent product over all primes:*

$$\zeta(s) = \prod_p \frac{1}{1 - p^{-s}}.$$

Proof. The idea is to expand each factor $(1 - p^{-s})^{-1}$ as a geometric series

$$\frac{1}{1 - p^{-s}} = \sum_{k=0}^{\infty} \frac{1}{p^{ks}}$$

and to multiply together the resulting series

$$\prod_p \frac{1}{1 - p^{-s}} = \sum \frac{1}{(p_1^{k_1} p_2^{k_2} \cdots \cdots \cdots p_r^{k_r})^s}.$$

We can write this as a sum

$$\sum_{n=1}^{\infty} \frac{a(n)}{n^s}$$

where $a(n)$ is the number of ways of expressing the integer n as a product of prime powers. By the Fundamental Theorem of Arithmetic 3, this can be done in one and only one way, i.e. $a(n) = 1$, which proves the product formula, once we check that everything is absolutely convergent if $\mathrm{Re}(s) > 1$.

As the above argument shows, the product formula is but a form of the Fundamental Theorem of Arithmetic.

8.1 Analytic Continuation and Functional Equation of $\zeta(s)$

To further explore the connection between the theory of primes and $\zeta(s)$, we will analytically continue $\zeta(s)$ to all values of s. We use the Gamma function given for $\mathrm{Re}(s) > 0$ by the integral representation

$$\Gamma(s) = \int_0^\infty e^{-t} t^s \frac{dt}{t}$$

to define the *completed zeta function* by

$$\zeta^*(s) := \pi^{-s/2} \Gamma(\frac{s}{2}) \zeta(s)$$

The basic fact about this variant of $\zeta(s)$ is

Theorem 12.   *1. The completed zeta function $\zeta^*(s)$ has a meromorphic continuation to the entire s-plane.*
   2. $\zeta^*(s)$ is analytic except for simple poles at $s = 0, 1$.
   3. It satisfies the functional equation

$$\zeta^*(s) = \zeta^*(1 - s)$$

As an immediate consequence of this fact, we observe that $\zeta^*(s)$ has no zeros outside the critical strip $0 \le \mathrm{Re}(s) \le 1$. This holds since $\Gamma(s)$ is never zero, and $\zeta(s)$ is analytic and nonzero in the region of convergence $\mathrm{Re}(s) > 1$, so that the completed zeta function $\zeta^*(s) \ne 0$ in $\mathrm{Re}(s) > 1$; by the functional equation, the same is true for the symmetric region $\mathrm{Re}(s) < 0$. Moreover, since $\Gamma(s)$ is analytic except for simple poles at $s = 0, -1, -2, \ldots$, $\zeta(s)$ is nonzero in $\mathrm{Re}(s) < 0$ except for simple zeros at the negative even integers $s = -2, -4, \ldots$ (to make up for the simple poles of $\Gamma(\frac{s}{2})$ at these points). These are called the *trivial zeros* of $\zeta(s)$; the nontrivial ones are the zeros of $\zeta^*(s)$ and as we have seen they all lie in the critical strip.

Proof. (Sketch) We start with the integral representation

$$\pi^{-s/2} \Gamma(\frac{s}{2}) \frac{1}{n^s} = \int_0^\infty e^{-\pi n^2 t} t^{s/2} \frac{dt}{t}$$

which shows that we have an integral representation of $\zeta^*(s)$ for $\mathrm{Re}(s) > 1$ as

$$\zeta^*(s) = \int_0^\infty \frac{\theta(t) - 1}{2} t^{s/2} \frac{dt}{t} \tag{8.1}$$

where the theta-function is given for $t > 0$ by

$$\theta(t) = \sum_{n=-\infty}^\infty e^{-\pi n^2 t}$$

By Poisson summation, $\theta(t)$ has a transformation formula

$$\theta\left(\frac{1}{t}\right) = \sqrt{t}\theta(t) \tag{8.2}$$

Breaking up the region of integration in the integral representation 8.1 to an integral over $(0, 1)$ and one over $(1, \infty)$, we change variables $t \mapsto 1/t$ to to transform the integral over (0.1) to one over $(1, \infty)$. We then use the transformation formula 8.2 for $\theta(t)$ to find after some manipulation that

$$\zeta^*(s) = -\frac{1}{s} - \frac{1}{1-s} + \int_1^\infty \frac{\theta(t)-1}{2}\left(t^{s/2} + t^{(1-s)/2}\right)\frac{dt}{t} \tag{8.3}$$

Since $\theta(t) - 1 = O(e^{-\pi t})$ as $t \to \infty$, the integral is absolutely convergent for all s and is therefore an entire function of s. Thus from (8.3) we get the meromorphic continuation, with the only poles being the simple ones at $s = 0, 1$. From the symmetry of (8.3) with respect to $s \mapsto 1 - s$ we get the functional equation.

8.2 Connecting the Primes and the Zeros of $\zeta(s)$

Riemann, in his seminal paper of 1858 [5], used $\zeta(s)$ to give a formula for $\pi(x)$ in terms of the zeros of $\zeta(s)$. His formula gives a clear understanding as to why $\mathrm{Li}(x)$ is the correct approximation to $\pi(x)$. Instead of a formula for $\pi(x)$, it is more convenient to give a formula for the weighted sum of prime powers $p^k \le x$, each prime power p^k weighted by the logarithm $\log p$ of the corresponding prime. One defines

$$\psi(x) := \sum_p \sum_{k:p^k \le x} \log p$$

The repetitions p^k for $k \ge 2$ give a contribution of the order of at most $\sqrt{x}$. The primes ($k = 1$) give a contribution which, if one believes Gauss' assertion that the density of primes near x is about $1/\log x$, is about x. Thus we expect (and the above argument is easily made rigorous) that the Prime Number Theorem is equivalent to the assertion that $\psi(x) \sim x$. This is made transparent by the formula (due to von Mangoldt)

$$\psi(x) = x - \sum_\rho \frac{x^\rho}{\rho} - \frac{\zeta'}{\zeta}(0) \tag{8.4}$$

where the sum is over all zeros ρ of $\zeta(s)$. Note that we cannot expect the formula to converge absolutely, since it would then define a *continuous* function of x, while $\psi(x)$ is a step function with jumps when $x = p^k$ is a prime power.

The contribution of the trivial zeros $\rho = -2, -4, -6, \ldots$ is easily summed to equal $\frac{1}{2} \log(1 - x^{-2})$ and is negligible. The constant term is $\zeta'/\zeta(0) = \log 2\pi$. The important part is the sum over the nontrivial zeros, which we expect to be of smaller order than x. It is thus crucial to understand the distribution of the zeros.

8.3 The Riemann Hypothesis

As noted in section 8.1, the nontrivial zeros of $\zeta(s)$ all lie in the critical strip $0 \leq \operatorname{Re}(s) \leq 1$. If ρ is a zero then by the functional equation $\zeta^*(s) = \zeta^*(1-s)$, so is $1 - \rho$, and since $\zeta(\bar{s}) = \overline{\zeta(s)}$ ($\bar{z}$ denoting complex conjugation), we get zeros at $\bar{\rho}$ and $1 - \bar{\rho}$ (the two symmetries $s \mapsto \bar{s}$ and $s \mapsto 1 - s$ coincide on the "critical line" $\operatorname{Re}(s) = 1/2$).

The first few zeros were computed by Riemann himself, and all lie on the critical line $\operatorname{Re}(s) = 1/2$. They are $\rho_n = 1/2 + it_n$ with $t_1 = 14.13\ldots, t_2 = 21.02\ldots, t_3 = 25.01\ldots$ etc. (by symmetry, we only need to consider positive t).

Riemann's Hypothesis (RH): *All nontrivial zeros of $\zeta(s)$ lie on the critical line* $\operatorname{Re}(s) = 1/2$.

The Riemann Hypothesis has been checked extensively and is widely believed to be true, though an explanation and proof are still missing to date. Its significance to the theory of primes is immense. For instance, we can use RH to explain the small size of the remainder term $\operatorname{Li}(x) - \pi(x)$ in Table 1. To see this, it suffices to show that $\psi(x) - x$ is small, and in fact we shall argue that it is of order at most $\sqrt{x} \log^2 x$. This is reasonable if we look at the formula for $\psi(x)$ in (8.4), which we will write as

$$\psi(x) = x - \sum \frac{x^{1/2+it_n}}{1/2 + it_n} + \ldots$$

where the sum is now only over the nontrivial zeros, the omitted terms being negligible. If we assume the t_n are *real*, so $|x^{1/2+it_n}| = \sqrt{x}$, it is tempting to then use the triangle inequality to deduce

$$|\psi(x) - x| \leq \sqrt{x} \sum \frac{1}{|1/2 + it_n|}$$

and so say that $\psi(x) - x$ is of order $\sqrt{x}$. The argument is not quite correct, as it transpires that the sum of absolute values diverges: $\sum 1/|1/2 + it_n| = \infty$. Nevertheless, this gives the essence of what is happening, and in fact taking more care and using more information on the distribution of zeros, one can show that $\psi(x) - x \ll \sqrt{x}\log^2 x$. This gives $\pi(x) - \mathrm{Li}(x) \ll \sqrt{x}\log x$ and so explains the observation regarding the size of the third column in Table 1.

The Riemann Hypothesis and its generalization (GRH) to other "L-functions" is one of the most important unsolved problems in Number Theory, and its validity has numerous implications. For instance, there are algorithms for *primality testing* of integers which are proved to require polynomial time (that is, testing if n is prime or not requires a number of operations polynomial in $\log n$), provided we assume GRH.

As to what was actually proved so far, the significant fact is that there are no zeros on the boundary of the critical strip: $\zeta(1 + it) \neq 0$, so $0 < \mathrm{Re}(\rho) < 1$ for all nontrivial zeros. This is enough to prove the Prime Number Theorem, as was done (independently) by Hadamard and de la Vallée Poussin in 1896. One has in fact a zero-free region near the boundary of the critical strip, whose width shrinks to zero as we go up. However, to date we do not have a proof that there is any strip of the form $\mathrm{Re}(s) > 1 - \delta$ in which there are no zeros, for any $\delta > 0$.

References

1. H. Davenport, *The higher arithmetic*, Seventh edition, Cambridge Univ. Press, Cambridge, 1999.
2. H.M. Edwards *Riemann's Zeta Function*, Academic Press 1974.
3. G.H. Hardy and E.M. Wright, *An introduction to the theory of numbers* (The Clarendon Press, Oxford University Press, New York, 1979).
4. M. Murty *Artin's conjecture for primitive roots* Math. Intelligencer **10** (1988), no. 4, 59–67.
5. B. Riemann *Über die Anzahl der Primzahlen unter einer gegebenen Größe*, Montasb. der Berliner Akad. (1858/60) 671–680, in *Gessamelte Mathematische Werke* 2nd edition, Teubner, Leipzig 1982 no. VII.

Mathematical Aspects of Quantum Maps

Mirko Degli Esposti[1] and Sandro Graffi

Dipartimento di Matematica, Università di Bologna, 40127 Bologna, Italy,
desposti@dm.unibo.it, graffi@dm.unibo.it

1 Introduction

In this contribution we will review the basic mathematical aspects of quantum maps. Here is a brief outline of the topics covered in this contribution. Most of the material comes from [9, 12, 13] and also [23, 24].

1. The space of the states and quantization of observables
2. Quantum dynamics over the torus
3. Quantized cat maps
4. The quantum baker's map and the sawtooth maps
5. Equidistribution of eigenfunctions
6. The period of $A \bmod N$ and an introduction to the Hecke operators
7. Value distribution of eigenfunctions
8. Equidistribution of Eigenfunctions for discontinuous maps

2 Quantization of Maps

Here we consider the quantization of measure preserving maps of the two-dimensional torus $\mathbb{T}^2 = \mathbb{R}^2/\mathbb{Z}^2$. While this quantization procedure was originally developed for linear automorphisms, that is elements of the modular group $SL(2, \mathbb{Z})$, we will see how it can be easily extended to an arbitrary (possibly discontinuous) symplectic maps, such as the baker's map or the sawtooth maps.

Here we will mostly concentrate on mathematical aspects related to quantization and semiclassical properties of toral maps. We refer to Andreas Knauf's contribution in these notes for what concerns results and definition out the (classical) dynamical system theory (see also Roberto Artuso's contribution).

Here we will restrict ourselves to work out, step by step, two specific models: the linear hyperbolic automorphisms and the baker's map.

Example 1 (Cat Map). The phase space of the system is the two dimensional torus $\mathbb{T}^2 = \mathbb{R}^2/\mathbb{Z}^2$, we use $(q, p) \in [0, 1] \times [0, 1]$ as coordinates and we denote by m the uniform Lebesgue measure.

We choose $A \in SL(2, \mathbb{Z})$ as [1]:

$$A = \begin{pmatrix} 2 & 1 \\ 3 & 2 \end{pmatrix}$$

The corresponding invertible transformation over the torus is give by:

$$(q, p) \longrightarrow (2q + p, 3q + 2p) \mod 1,$$

with inverse induced by $A^{-1} = \begin{pmatrix} 2 & -3 \\ -1 & 2 \end{pmatrix}$, i.e.

$$(q, p) \longrightarrow (2q - 3p, -q + 2p) \mod 1.$$

A has determinant 1, namely the transformation is measure preserving: if $Q \subset \mathbb{T}^2$, then $m(Q) = m(A(Q)) = m(A^{-1}(Q))$. The eigenvalues of A are $(\lambda, \lambda^{-1}) = (2+\sqrt{3}, 2-\sqrt{3})$, and the corresponding expanding and contracting eigenspaces through the origin are given by $v = (-\sqrt{3}, 1)$ and $w = (\sqrt{3}, 1)$ respectively.

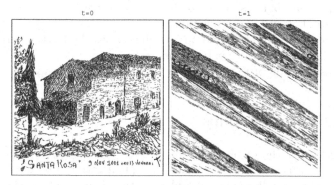

Fig. 1. Action of the map A at times $t = 0$ and $t = 1$

As it is well known, the irrational slope of the two directions implies that stable and unstable (linear) manifolds are densely distributed over the torus. This combination of expanding/contracting and folding back to the compact torus is responsible of the well known strongly chaotic properties of this kind of maps (see again A. Knauf's contribution). Here we just show in Fig. 1 and Fig. 2 how the map acts.

---

[1] In the literature, the so called "Arnold's cat map" coincide with the map $A = \begin{pmatrix} 2 & 1 \\ 1 & 1 \end{pmatrix}$. For reasons that will be clear when we will define the corresponding quantum propagator, we use instead the one previously defined. The two systems are completely equivalent, but with this choice we can avoid an unnecessary extra formalism.

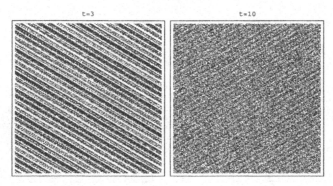

Fig. 2. Action of the map A at times $t = 3$ and $t = 10$.

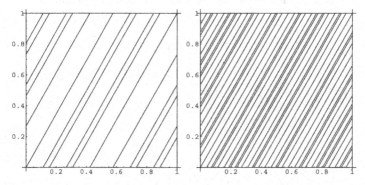

Fig. 3. Increasing portion of the expanding manifold passing through the origin.

Example 2 (Baker's Map). The baker's map is another canonical and fundamental example in the realm of two dimensional discrete systems [1]. Let

$$A = \begin{pmatrix} 2 & 0 \\ 0 & \frac{1}{2} \end{pmatrix}. \tag{1}$$

The baker's map is a discontinuous map B on the torus, defined as follows [1]: $(x = (q,p) \in \mathbb{T}^2)$

$$B(q,p) = \begin{cases} Ax, & q \in [0, 1/2[, \\ Ax + (-1, +1/2), & q \in]1/2, 1]. \end{cases} \tag{2}$$

We refer to [1] for the ergodic and topological properties of the map B. Here we just reduce ourselves to show in Fig. 4, Fig. 5 and Fig. 6 some iterates of the baker's map, starting from a cloud of initial conditions confined in the strip $0 \le q \le 1/2$. The "cutting-expanding-stretching and folding" effect of the map is quite clear. A better understanding of its dynamics can be achieved by a natural coding of the orbits by *bi-infinite* words out of a two symbols alphabet. More precisely, given a point $x = (q,p)$ in phase space $(0 \le q, p < 1)$, consider the binary expansion of each coordinate:

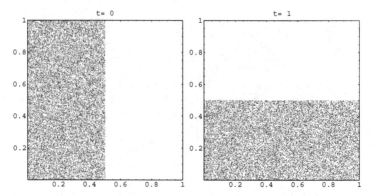

Fig. 4. Action of the baker's map B at times $t = 0$ and $t = 1$

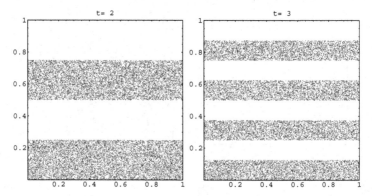

Fig. 5. Action of the baker's map B at times $t = 2$ and $t = 3$

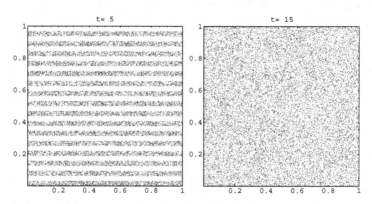

Fig. 6. Action of the baker's map B at times $t = 5$ and $t = 15$

$$q = .\epsilon_0\epsilon_1 \cdots, \quad p = .\tau_1\tau_2\cdots, \quad \epsilon_j, \tau_j \in \{0,1\}.$$

Almost[2] all points in phase space can be uniquely identified by the infinite word (where the origin · is identified):

$$\omega = \omega(q,p) = \cdots \tau_3\tau_2\tau_1 \cdot \epsilon_0\epsilon_1 \cdots.$$

The baker's map can now be written as:

$$B(q,p) = (2q - \epsilon_0, (p + \epsilon_0)/2).$$

The baker's map action on symbols ω turns out to be just the left-shift operation:

$$\omega(B(q,p)) = \cdots \tau_3\tau_2\tau_1\epsilon_0 \cdot \epsilon_1\epsilon_2 \cdots.$$

Conversely, prescribed sequences with a given structure can be used to construct (a family of) orbits with a given symmetry (see [37]). In particular, the map is area-preserving, uniformly hyperbolic with a constant Lyapunov exponent $\lambda = \log 2$, the vertical (horizontal) lines $q = q_0$ ($p = p_0$) represents the stable (unstable) foliations. The discontinuous dependence upon ϵ_0 is the only "cutting" effect responsible for the discontinuities of the map.

Referring again to [37] for more details on the subject, here we just point out, as it can be realized looking at the binary coding, that there are two natural symmetries commuting with the map. One is related to *time-reversal*, the other to the *parity* operator $P(q,p) = (1 - q, 1 - p)$. These observations will play a rule in deciding which quantum propagator should be associated to the classical map B.

Now we go back to the problem of finding a quantum counterpart of these systems. The first approach to quantization of toral maps was presented in [3]. We will follow here an approach by means of representation theory which was developed in [12, 13]. See also [2, 11, 21, 40] for other equivalent approaches.

In Bäcker's contribution and in references therein, the reader can find another equivalent approach based on generating functions. Here we prefer to stress the one which has a more algebraic character and allows to explore the many connections with number theory. Following the general approach to quantization, we divide our presentation in two parts: the kinematic and the dynamics.

In the first part we will define the suitable Hilbert space $\mathcal{H}_\hbar$ (for very specific values of the Planck's constant $\hbar$) of states of the quantum system and an algebra of operators on the space obtained through a quantization procedure starting from smooth classical functions on phase space (*algebra of quantum observables*). In our setup, the classical phase space is the torus $\mathbb{T}^2$ and the classical observable are given by $C^\infty(\mathbb{T}^2)$ functions. As we will see, the Planck constant $\hbar$ is restricted to be an inverse of an integer. More

---

[2] Up to a set of Lebesgue measure zero

precisely, doing quantum mechanics on the torus restricts the possible values of $\hbar$ to the values which satisfy the integer condition $2\pi\hbar N = 1$, for some $N \in \mathbb{N}$. The classical limit coincides with the limit $N \to \infty$, $N \in \mathbb{N}$.

The space of quantum states $\mathcal{H}_\hbar$ will be identified with the finite dimensional Hilbert space $\mathcal{H}_N = L^2(\mathbb{Z}_N)$, where $\mathbb{Z}_N$ is the set $\{0, \ldots, N-1\}$. Quantum observables are constructed by associating to each $f \in C^\infty$ a suitable operator $Op_\hbar(f)$ on $\mathcal{H}_N$, i.e. an $N \times N$ matrix. This procedure (*discrete Weyl quantization*) will satisfy, in the classical limit $N \to \infty$, the *usual* relations between the quantum commutator and the classical Poisson brackets relations induced by the symplectic structure. As we will see in details later on, $\forall g, f \in C^\infty(\mathbb{T}^2)$:

$$\frac{2\pi i}{\hbar}\left[Op_\hbar(f), Op_\hbar(g)\right] \to_{\hbar \to 0} Op_\hbar(\{f, g\}).$$

Finally, the dynamical part will be implemented by associating to each symplectic map $F : \mathbb{T}^2 \to \mathbb{T}^2$ a suitable unitary operator $U_N(F)$ acting on $\mathcal{H}_N$. We will first discuss this *natural* quantization procedure for elements of the group of linear automorphisms $SL(2, \mathbb{Z})$ (Example 1). We will then discuss the quantization of certain discontinuous maps, such as the baker's map (Example 2) and the so called sawtooth maps.

We now start with the kinematic part.

2.1 Kinematics: The Space of States and Quantization of Observables

Quantization of the Torus

A canonical way of setting up the quantum mechanics over the torus is to adapt the "usual" quantization procedure on $\mathbb{R}^2$, as obtained through the study of the representation of the so called "Heisenberg group" $\mathbb{H}_n(\mathbb{R})$. For the sake of completeness, we have included a small review of this in Appendix A.

As we will see, one can naturally adapt the above canonical quantization procedure to the case in which the phase space is the torus $\mathbb{T}^2$ and the underlying Heisenberg group is the subgroup of $\mathbb{H}_n(\mathbb{R})$ denoted by $\mathbb{H}_n(\mathbb{Z})$.

Before going into details concerning the discrete Heisenberg group and its representations, let us review a very simple argument which shows us that the Hilbert space of quantum states over the torus must be finite dimensional. We will then focus on a more formal and general approach.

As Hilbert space of states, we consider distributions $\psi(q)$ on the real line $\mathbb{R}$ which are periodic in both the position and the momentum representation, the last one being defined as: ($\hbar > 0$)

$$\mathcal{F}_N\psi(p) = \frac{1}{\sqrt{2\pi\hbar}}\int_{-\infty}^{+\infty}\psi(q)\, e^{-i\hbar^{-1}qp}\, dq\, .$$

By imposing [3]

$$\psi(q+1) = \psi(q), \qquad \mathcal{F}_N\psi(p+1) = \mathcal{F}_N\psi(p),$$

the periodicity in the position representation yields:

$$\psi(q) = \sum_{n\in\mathbb{Z}} c_n\, e^{2\pi i n q}.$$

If we now apply $\mathcal{F}_N$, in the momentum representation we get:

$$\mathcal{F}_N\Psi(q) = \sqrt{2\pi\hbar} \sum_{n\in\mathbb{Z}} c_n\, \delta(p - 2\pi n\hbar).$$

It is now easy to see that periodicity in both q and p coordinate implies:

$$2\pi\hbar N = 1, \quad N \in \mathbb{N},$$

and also:

$$c_{n+N} = c_n.$$

Let us now turn to the more general and formal approach. We remark first that the 2-dimensional torus can be considered a submanifold of $\mathbb{C}^2$, under the usual covering map

$$\pi : \mathbb{R}^2 \longrightarrow \mathbb{T}^2 : \ \pi(q,p) = (\eta, \xi) = (e^{2\pi i q}, e^{2\pi i p}).$$

The action of π on the symplectic form $\omega = dp \wedge dq$ yields

$$\Omega = \frac{d\eta \wedge d\zeta}{(2\pi i)^2 \zeta \eta} : \quad \pi_*\Omega = \omega.$$

The symplectic form Ω induces a canonical Lie structure on the space of smooth functions defined on the torus, namely

$$\{f(\eta,\zeta), g(\eta,\zeta)\} \equiv (2\pi i)^2 \zeta\eta[\partial_\eta f \partial_\zeta g - \partial_\eta g \partial_\zeta f] = \Omega^{-1}(df, dg).$$

Writing:

$$\xi^{n_2}\eta^{n_1} \equiv \chi(n) : \ n = (n_1, n_2),$$

we immediately have

$$\{\chi(m), \chi(n)\} = (2\pi i)^2 \omega(n, m)\, \chi(n+m),$$

where, of course, $\omega(n,m) = n_1 m_2 - n_2 m_1$.
If f, g have the form

---

[3] Actually, in the general treatment, we will require the equality up to a phase, i.e. we will impose in general quasi-periodic conditions at the boundary

$$f = \sum_{n \in \mathbb{Z}^2} f_n \chi(n) \quad ; \quad g = \sum_{n \in \mathbb{Z}^2} g_n \chi(n),$$

then their Poisson bracket is

$$\{f, g\} = \sum_{n,m} f_n g_m (2\pi i)^2 \omega(n, m) \chi(n + m).$$

In order to canonically quantize the observables defined on the torus according to the above procedure, the natural objects to look are the ones obtained through the quotient by $\mathbb{Z}^2$ of the unique infinite dimensional unitary representation of the standard Heisenberg group $\tilde{\mathbb{A}}_h$ over the plane (see Appendix A):

Definition 1. – $\forall h \in \mathbb{R} \backslash 0$, the discrete Heisenberg group $\mathbb{H}_1 := \mathbb{H}_1(\mathbb{Z})$ is the subgroup topologically equivalent to $\mathbb{Z}^2 \times \mathbb{R}$ with group law

$$(n, t)(m, s) = (n + m, t + s + \frac{1}{2}\omega(n, m))$$

– The discrete Heisenberg algebra $\mathbb{A}_h$ is the subalgebra of $\tilde{\mathbb{A}}_h$ defined as the unitary *-algebra over $\mathbb{C}$ generated by the group

$$T_h = \{T(n)\}_{n \in \mathbb{Z}^2}$$

where (we use the abbreviation $T_{\hbar} = T$)

$$T(n)^* = T(-n) \tag{3}$$
$$T(n)T(m) = e^{i\pi h \omega(n,m)} T(n + m) \tag{4}$$

and the canonical quantization is obtained upon classification of all irreducible representations of $\mathbb{A}_h$, defined abstractly by the relations (3) and (4), into the unitary operators acting in the Hilbert spaces $L^2(S^1; \lambda)$, for some measure λ on S^1 to be determined.

Before turning to that, we compute the commutators of the algebra and of the observables. From (4), we immediately get the expression for the commutator of the algebra

$$[T(n), T(m)] = 2i \sin(\pi h \omega(n, m)) T(n + m).$$

To each $f = \sum_{n \in \mathbb{Z}^2} f_n \chi(n)$ we associate the element of the algebra, denoted by $Op_{\hbar}(f)$, formally obtained replacing $\chi(n)$ by $T(n)$ into the Fourier expansion. Following [12], we now go back to describe the irreducible representations of the discrete Heisenberg group.

To this end, let us first consider the generators of the algebra, defined as:

$$t_1 := T(1, 0); \quad t_2 := T(0, 1).$$

In fact, if $n = (n_1, n_2)$ we have:

$$T(n) = \exp\left[\pi i \frac{n_1 n_2}{N}\right] t_2^{n_2} t_1^{n_1}, \tag{5}$$

$$t_2 t_1 = e^{-2i\pi h} t_1 t_2. \tag{6}$$

If $h = 1/N$ then t_1^{N} and t_2^{N} are the generators of the center and each one has to be mapped into a unitary scalar multiple of the identity by any irreducible representation. The two corresponding phases define the representation which, as we will see soon, are always finite dimensional. In other words, the unique infinite dimensional unitary representation of the standard Heisenberg group $\tilde{A}_h$ splits into a direct integral (over the two torus) of finite dimensional, non equivalent, unitary representations.

In order to describe these representations consider once again the N dimensional Hilbert space $L^2(S^1, \mu_N)$, where, $\forall h = N^{-1}$, $\mu_N(x)$ is the atomic measure on the circle defined by :

$$\mu_N(x) := \frac{1}{N} \sum_{k_0}^{N-1} \delta(x - \frac{k}{N}).$$

The vectors $|k\rangle = \Psi_k(x) = \delta_{k/N}^x$ for $k = 0, 1, .., N-1$ are a basis of the Hilbert space $L^2(S^1, \mu_N)$, with the inner product between $\psi, \phi \in L^2(S^1, \mu_N)$ given by:

$$\langle \psi, \phi \rangle := \frac{1}{N} \sum_{l=0}^{N-1} \overline{\psi}(\frac{l}{N}) \phi(\frac{l}{N}) = \int_{S^1} \overline{\psi}(x) \phi(x) \, d\mu_N(x).$$

The action of the Fourier transformation on $L^2(S^1, \mu_N)$ is:

$$(\mathcal{F}_N \psi)_m := \frac{1}{\sqrt{N}} \sum_{n=0}^{N-1} \exp\left(\frac{2i\pi mn}{N}\right) \psi_n.$$

Writing

$$\mathbb{Q}_N = \{0, 1/N, 2/N, ..., (N-1)/N\}$$

and

$$\mathbb{Z}_N := \mathbb{Z}/N\mathbb{Z} = \{0, 1, 2, \ldots, N-1\},$$

we can identify $L^2(S^1, \mu_N)$ with $L^2(\mathbb{Z}_N, \mu_N)$ where $\psi \in L^2(\mathbb{Z}_N, \mu_N)$ is a vector in $\mathbb{C}^N$: $\psi = (\psi_0, \ldots, \psi_{N-1})$.

Now, for any fixed $\theta = (\theta_1, \theta_2) \in \mathbb{T}^2$, we define the representations of our algebra on $L^2(S^1, \mu_N)$ by specifying their action on the generators (identified with the corresponding matrices):

$$t_1|l\rangle = \exp(\frac{2i\pi(\theta_1 + l)}{N})|l\rangle, \qquad t_1^{N} = e^{2\pi i\theta_1} \cdot Id \tag{7}$$

$$t_2|l\rangle = \exp(\frac{2i\pi\theta_2}{N})|l+1\rangle, \qquad t_2^{N} = e^{2\pi i\theta_2} \cdot Id. \tag{8}$$

It is not difficult to see ([12]) that these representations are irreducible, non equivalent for different values of θ and moreover that they are the only possible ones.

For any fixed $2\pi\hbar = 1/N$ and $\theta \in \mathbb{T}^2$ we denote by $T_{N,\theta}$ the representation of $T(n)$ just constructed:

$$T_{N,\theta}(n) = e^{\left[\frac{\pi i n_1 n_2}{N}\right]} t_2{}^{n_2} t_1{}^{n_1}.$$

Because our main interests concern the asymptotic properties of eigenfunctions and eigenvalues for quantized map in the classical limit $N \to \infty$, we will often assume periodic boundary conditions for the quantum states, i.e. $\theta = (0,0)$.

The generator t_1 and t_2 correspond to the exponential of the "usual" position and momentum operators

$$\hat{q}\,\psi(q) := q\psi(q), \qquad \hat{p}\,\psi(q) := -i\hbar\frac{d\psi}{dq}(q).$$

The usual commutation relation $[\hat{q}, \hat{p}] = i\hbar$ is then equivalent to the relation (6). The action of $T_N(n)$ on a wave-function $\psi \in L^2(\mathbb{Z}_N)$ is

$$T_N(n)\psi(q) := e^{\frac{i\pi n_1 n_2}{N}} e^{\frac{2\pi i n_2 q}{N}} \psi(q + n_1), \qquad q \in \mathbb{Z}_N.$$

These operators are clearly of period $2N$ in n:

$$T_N(n + 2N \cdot m) = T_N(n), \quad \forall n, m \in \mathbb{Z}^2.$$

We can now construct quantum observables: for any smooth classical observable $f \in C^\infty(\mathbb{T}^2)$ with Fourier expansion: $(\chi(n) = e^{2\pi I(n_1 q + n_2 p)})$

$$f(q,p) = \sum_{n \in \mathbb{Z}^2} f_n \chi(n)$$

we define its quantization $Op_N(f)$ as

$$Op_N(f) := \sum_{n \in \mathbb{Z}^2} f_n T_N(n).$$

We can finally summarize what we have achieved by doing quantum mechanics over the torus as phase space:

(i) *The Planck's constant $\hbar$ can only takes values $2\pi\hbar N = 1$, $N \in \mathbb{N}$*
(ii) Given $h = 1/N$, $N \in \mathbb{N}$, the *space of states:* $\mathcal{H}_N$ is finite dimensional, of dimension N. We identify $\mathcal{H}_N$ with the N-dimensional vector space $L^2(\mathbb{Z}/N\mathbb{Z})$, with inner product

$$\langle \Psi, \Phi \rangle = \frac{1}{N}\sum_{k=0} \bar{\Psi}(k)\Phi(k)$$

(iii) *quantization of the observables*: there is a well defined map

$$f(q,p) \in C^\infty(\mathbb{T}^2) \to Op_\hbar(f)$$

which allows to associate a suitable operator on $\mathcal{H}_N$ to any smooth observable on phase space. This map does in fact implement the usual commutation relations and it can be seen as the discrete analog of the usual Weyl quantization which associate a zero-order pseudodifferential operator to any given bounded symbol.

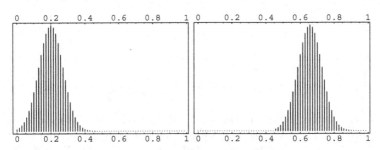

Fig. 7. Action of the phase space translations $T(n,m)$. *Left*: plot of $|\psi|^2$ versus k/N with $N = 70$. *Right*: plot of $|T(32,0)\psi|^2$.

2.2 Quantum Dynamics

We now focus on the problem of defining a suitable quantum operator corresponding to a given are preserving dynamics. We recall that there exists a general construction which allows to assign to a given area preserving map $F : \mathbb{T}^2 \to \mathbb{T}^2$ a suitable $N \times N$ unitary matrix $U_N(F)$ acting on $\mathcal{H}_N = L^2(\mathbb{Z}_N)$ and implementing the quantum evolution on the space of states.

In these notes we will mostly focus on the algebraic case given by dynamics induced by automorphisms $A \in SL(2,\mathbb{Z})$ (see Bäcker's contribution for "perturbed cat maps") and the baker's map. It should not come as a surprise that number theory will play a crucial rule in exploring the asymptotic properties (such as equidistribution of eigenfunctions or spectral statistics (see also Bäcker's contribution)) of the $U_N(A)$'s.

We refer the reader to [19, 20] for more detailed results concerning the properties and the definitions of the quantum propagator associated to generic Anosov toral map.

Quantized Cat Maps

The modular group $SL(2,\mathbb{Z})$ acts by automorphisms on the Heisenberg group by $A \cdot (x,s) = (A^T x, s)$, where $x = (q,p)$, $s \in \mathbb{R}$, $A \in SL(2,\mathbb{Z})$ and where A^T

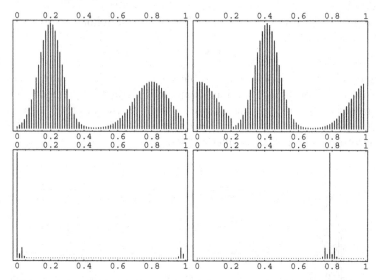

Fig. 8. The action of $T(15, 15)$ in both configuration (*top*) and momentum (*bottom*) representation. *Top*: plot of $|\psi|^2$ versus k/N with $N = 70$ (*left*) and of $|T(15, 15)\psi|^2$ (*right*). *Top*: plot of $|\mathcal{F}_N(\psi)|^2$ versus k/N with $N = 70$ (*left*) and of $|\mathcal{F}_N(T(15, 15)\psi)|^2$ (*right*).

denotes the transpose matrix (sometimes we might use xA instead of $A^T x$). In particular for any given representation $T_{N,\theta}(n)$, we get a new representation defined by $\rho_A(n) = T_{N,\theta} \circ A^T(n) = T_{N,\theta}(A^T n) : \forall n \in \mathbb{Z}^2$. By looking at the action on the central character, it is easy to see that ρ_A is again an irreducible representation, and thus unitarily equivalent to one of the previous. The unitary intertwining operator realizes the "commutativity between classical evolution and quantization" and thus represents the quantum propagator. The proof of the following is left as an exercise [12].

Theorem 1. *Let* $A = \begin{pmatrix} a & b \\ c & d \end{pmatrix} \in SL(2, \mathbb{Z})$. *Then:*

(i) $\rho_A(n) = T_{N,\theta}(A^T n)$ *is again an irreducible representation of* $\mathbb{A}_h$;
(ii) $\forall \theta \in \mathbb{T}^2$, *there exists a unitary operator* $U_A(\theta)$ *such that,* $\forall n \in \mathbb{Z}^2$:

$$T_{N,\theta}(A^T n) = U_A(\theta)^{-1} \circ T_{N,\theta}(n) \circ U_A(\theta)$$

where

$$\varphi_A(\theta) = A(\theta) + \frac{1}{2}\begin{pmatrix} abN \\ cdN \end{pmatrix} \quad mod\ 1.$$

For the sake of simplicity and because of our interest in semiclassical results, we will now restrict our analysis to a particular subgroup of automorphisms. This allows us to consider only periodic wave-functions, i.e. from now on we take $\theta = (0,0)$ (see [11, 12, 19, 40] for the general quantization procedure). We start by defining the theta group $\Gamma_\theta(2N)$: for any given $N \in \mathbb{N}$ we let

$$\Gamma_\theta(2N) = \left\{ \begin{pmatrix} a & b \\ c & d \end{pmatrix} \in SL(2, \mathbb{Z}_N) : ab \equiv cd \equiv 0 \bmod 2 \right\}.$$

From Theorem 1, it is easy to see that the representation with $\theta = (0,0)$ is preserved by the action of the elements of $\Gamma_\theta(2N)$. In particular the intertwining operators can be chosen to preserve also the multiplicative structure [24, 30]:

Theorem 2. *There exists a map* $A \in \Gamma_\theta(2N) :\to U_N(A)$ *into the* $N \times N$ *unitary matrices, such that*

(i)

$$U_N(A)^{-1} T_N(n) U_N(A) = T_N(A^t n) \qquad \forall n \in \mathbb{Z}^2$$

In particular, $\forall f \in C^\infty(\mathbb{T}^2)$

$$Op_N(f \circ A) = U_N(A)^{-1} Op_N(f) U_N(A)$$

(ii) $\forall A, B \in \Gamma_\theta(2N)$

$$U_N(AB) = U_N(A) \circ U_N(B)$$

We now discuss the dependence of our construction from the arithmetic properties of N [24]. The Chinese remainder Theorem allows us to reduce all cases to prime powers: factor $N = \prod_p p^{k_p} = 2^k \prod_{p>2} p^{k_p} = 2^k M$, M odd. The map $x \to \{x \bmod p^{k_p}\}_p$ induces the isomorphism

$$\mathbb{Z}/N\mathbb{Z} \equiv \prod_p \mathbb{Z}/p^{k_p}\mathbb{Z}.$$

The inverse map is given by:

$$\{x_p \bmod p^{k_p}\}_p \to \sum_p \frac{N}{p^{k_p}} r_p x_p \bmod N,$$

where r_p is the inverse of N/p^{k_p} modulo p_p^k.

Note that the previous map induces also the bijection

$$L^2(\mathbb{Z}_N) \equiv \otimes_p L^2(\mathbb{Z}_{p^{k_p}}).$$

It is now easy to see that the previous quantization preserves the decomposition into prime powers and it allows to reduce the proof of Theorem 2 only to values of N which are powers of primes: if $\psi \in L^2(\mathbb{Z}_N)$ is decomposable as $\psi = \otimes_p \psi_p$, $\psi_p \in L^2(\mathbb{Z}_{p^{k_p}})$, then

(i)

$$T_N(n) = \otimes T_{p^{k_p}}(n),$$

namely

$$T_N(n)\psi(q) = \prod_p T_{p^{k_p}}(n)\psi(q \bmod p^{k_p})$$

(ii) If the propagators $U_{p^{k_p}}(A)$ have been already defined and if they satisfy

$$U_{p^{k_p}}(A)^{-1}T_N(n)U_{p^{k_p}}(A) = T_N(A^T n), \qquad n \in \mathbb{Z}^2,$$

then the propagator

$$U_N(A) = \otimes_p U_{p^{k_p}}(A) \qquad (9)$$

does in turn satisfy the first statement of the previous Theorem, i.e.

$$U_N(A)^{-1}T_N(n)U_N(A) = T_N(A^t n) \qquad \forall n \in \mathbb{Z}^2$$

Remark 1. As discussed in [24], we use this procedure to define $U_N(A)$, that is we choose a phase, so that $U_N(\cdot)$ is an honest (i.e. not just projective) representation of a subgroup $\Gamma(4, 2N) \subset SL(2, \mathbb{Z}_N)$. From the factorization property (9), it follows that it is enough to show that the map $A \to U_{p^{k_p}}(A)$ is a representation of $SL(2, \mathbb{Z}_{p^{k_p}})$ when $p > 2$ is odd, and of $\Gamma(4, 2^k)$ if $N = 2^{k-1}M$ is even.

We describe now how to define U_{p^k} when p is an odd prime and we refer to [24] for the case $N = 2^k$. We start with a classical result:

Proposition 1. *If p is an odd prime then the group $SL(2, \mathbb{Z}_{p^k})$ is generated by the matrices*

$$\begin{pmatrix} 1 & b \\ 0 & 1 \end{pmatrix}, \begin{pmatrix} t & 0 \\ 0 & t^{-1} \end{pmatrix}, \begin{pmatrix} 0 & 1 \\ -1 & 0 \end{pmatrix}$$

We now specify U_{p^k} for such matrices and verify the previous relations.

Before giving the formulas, we let $U = U_{p^{k_p}}$, $e(x) = e^{2\pi i x}$, given also $a, r \in \mathbb{Z}$ we denote with $S_r(a, p^k)$ the normalized Gauss sum:

$$S_r(a, p^k) = \frac{1}{\sqrt{p^k}} \sum_{x=0}^{p^k-1} e(-\frac{rax^2}{p^k}).$$

Moreover, we let:

$$\Lambda_{r,p^k}(t) = \frac{S_r(t, p^k)}{S_r(1, p^k)}.$$

Because we assume p a odd prime, $S_r(a, p^k)$ is one of the 4th roots of unity, and we have:

$$\Lambda_{r,p^k}(t) = \frac{S_r(t, p^k)}{S_r(1, p^k)} = \left(\frac{t}{p}\right)^k,$$

where $\left(\frac{t}{p}\right)$ is the Legendre symbol over $\mathbb{Z}_p$, namely the multiplicative character of order 2 given by $\left(\frac{t}{p}\right) = 1$ if $t = x^2 \bmod p$ and $\left(\frac{t}{p}\right) = -1$ otherwise (see also Rudnick's contribution). Finally, $S_r(-1, p^k) = 1$ if k is even and

$S_r(-1, p^k) = \epsilon(p) \left(\frac{r}{p}\right)$ if k is odd, where $\epsilon(p) = 1$ if $p = 1$ mod 4 and $\epsilon(p) = i$ if $p = 3$ mod 4.

We are now ready to define the quantum propagator for the generators of $SL(2, \mathbb{Z}_{p^k})$:

(i) If $A = \begin{pmatrix} 1 & b \\ 0 & 1 \end{pmatrix}$ then

$$U(A)\psi(q) = e\left(\frac{rbx^2}{p^k}\right)\psi(x)$$

(ii) If $\begin{pmatrix} t & 0 \\ 0 & t^{-1} \end{pmatrix}$ then

$$U(A)\psi(q) = \Lambda_{r,p^k}(t)\,\psi(tx)$$

(iii) If $A = \begin{pmatrix} 0 & 1 \\ -1 & 0 \end{pmatrix}$ then

$$U(A)\psi(q) = S_r(-1, p^k)\frac{1}{\sqrt{p^k}} \sum_{y \bmod p^k} e\left(\frac{2rxy}{p^k}\right)\psi(y)$$

and it is now not difficult to see that these unitary matrices does in fact satisfy the properties stated in Theorem 2.

Example 3 (Cat Map Quantum Propagator). The 2×2 matrices of the form $A = \begin{pmatrix} 2m & 1 \\ 4m^2 - 1 & 2m \end{pmatrix}$, $\forall m \in \mathbb{N}$, all belongs to $\Gamma_\theta(2N)$.

In this case, up to a phase:

$$U_N(A)_{jk} = \frac{1}{\sqrt{N}} \exp \frac{2\pi i}{N} \left[mk^2 - kj + mj^2\right].$$

In Example 1 we have $m = 1$, and if $N = 5$:

$$U_N(A) = \begin{pmatrix} \frac{1}{\sqrt{5}} & \frac{e^{\frac{2i}{5}\pi}}{\sqrt{5}} & \frac{e^{\frac{-2i}{5}\pi}}{\sqrt{5}} & \frac{e^{-\frac{2i}{5}\pi}}{\sqrt{5}} & \frac{e^{\frac{2i}{5}\pi}}{\sqrt{5}} \\ \frac{e^{\frac{2i}{5}\pi}}{\sqrt{5}} & \frac{e^{\frac{2i}{5}\pi}}{\sqrt{5}} & \frac{e^{\frac{-4i}{5}\pi}}{\sqrt{5}} & \frac{e^{\frac{4i}{5}\pi}}{\sqrt{5}} & \frac{e^{-\frac{4i}{5}\pi}}{\sqrt{5}} \\ \frac{e^{-\frac{2i}{5}\pi}}{\sqrt{5}} & \frac{e^{-\frac{4i}{5}\pi}}{\sqrt{5}} & \frac{e^{-\frac{2i}{5}\pi}}{\sqrt{5}} & \frac{e^{\frac{4i}{5}\pi}}{\sqrt{5}} & \frac{e^{\frac{4i}{5}\pi}}{\sqrt{5}} \\ \frac{e^{-\frac{2i}{5}\pi}}{\sqrt{5}} & \frac{e^{\frac{4i}{5}\pi}}{\sqrt{5}} & \frac{e^{\frac{4i}{5}\pi}}{\sqrt{5}} & \frac{e^{-\frac{2i}{5}\pi}}{\sqrt{5}} & \frac{e^{\frac{-4i}{5}\pi}}{\sqrt{5}} \\ \frac{e^{\frac{2i}{5}\pi}}{\sqrt{5}} & \frac{e^{\frac{4i}{5}\pi}}{\sqrt{5}} & \frac{e^{\frac{4i}{5}\pi}}{\sqrt{5}} & \frac{e^{-\frac{4i}{5}\pi}}{\sqrt{5}} & \frac{e^{\frac{2i}{5}\pi}}{\sqrt{5}} \end{pmatrix}$$

See Fig. 9 for an example of quantum evolution.

We now turn to review briefly the quantization procedure for a class of discontinuous and not completely algebraic dynamical systems, namely the so called baker's map [4, 36, 37] and the sawtooth maps. Despite of this, they

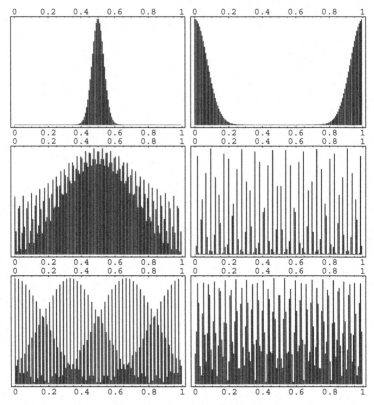

Fig. 9. Evolution of a given initial state ψ (*top left*) under the action of $U_N(A)^t$. $N = 141$ and A as in Example 1. $t = 0, 1, 3, 4, 5, 10$ from *top left to bottom right*.

will nevertheless retain some of the rigid features met in the arithmetic cat map. As we will see, the presence of discontinuities makes the model quite "generic", at least for what concerns the statistical properties of eigenfunctions and eigenvalues, meanwhile the presence of a mild algebraic structure and a quite simple coding for the classical orbits allows to push forward several semiclassical results.

The Baker's Map

Here we consider the measure preserving discontinuous map defined in Example 2. Its quantization and some of its semiclassical features has been studied quite deeply in [36, 37]. Here we just briefly recall the expression of the quantum propagator (see also [9] and references therein). Let us look first at the action of the classical maps characters $\chi(n, m) = e^{2\pi i[nq - mp]}$.

It follows from the definition that ($\forall n, r \in \mathbb{Z}$)

$$\chi(n, 2r) \circ B = \chi(2n, r), \tag{10}$$

$$\chi(n, 2r+1) \circ B = \begin{cases} e^{-i\pi p}\chi(2n, r) & \text{if } q \in [0, 1/2[\\ -e^{-i\pi p}\chi(2n, r) & \text{if } q \in [1/2, 1[. \end{cases} \tag{11}$$

With the assumption $N = 0 \bmod 2$, the quantization of the baker's map B turns out to be given by the following unitary operator in the canonical q-basis $\mathbf{e}_k$, $k = 0, \ldots, N-1$ [4,11,36,37]:

$$U_N(B) = \mathcal{F}_N^{-1} \circ \begin{pmatrix} \mathcal{F}_{N/2} & 0 \\ 0 & \mathcal{F}_{N/2} \end{pmatrix} = \mathcal{F}_N^{-1} \circ Q_B. \tag{12}$$

$U_N(B)$ acts on $\mathcal{H}_N = \mathcal{H}_{N,L} \oplus \mathcal{H}_{N,R}$. Here, with $\mathbb{Z}_N = \{0, \ldots, \frac{N}{2} - 1\} \cup \{\frac{N}{2}, \ldots, N-1\} = L \cup R$:

$$\mathcal{H}_{N,L} = \operatorname{span}_{\mathbb{C}} \{\mathbf{e}_\ell\}_{\ell \in L}, \qquad \mathcal{H}_{N,R} = \operatorname{span}_{\mathbb{C}} \{\mathbf{e}_\ell\}_{\ell \in R}. \tag{13}$$

With this notations:

$$\left(\mathcal{F}_{N/2}\right)_{k,\ell} = \sqrt{\frac{2}{N}} e^{-\frac{2\pi i}{\frac{N}{2}} k\ell} \qquad k, \ell \in \{0, \ldots, \frac{N}{2} - 1\}. \tag{14}$$

Given $\ell \in \mathbb{Z}_N$, let the "strip" S_ℓ of ℓ be defined by $S_\ell = L$ ($S_\ell = R$) if $\ell \in L$ ($\ell \in R$). Then $\forall \ell, k \in \mathbb{Z}_N$

$$\langle \mathbf{e}_k, Q_B \mathbf{e}_\ell \rangle = \begin{cases} \sqrt{\frac{2}{N}} e^{-\frac{4i\pi}{N}\ell k} & \text{if } S_\ell = S_k \\ 0 & \text{if } S_\ell \neq S_k \end{cases} \tag{15}$$

It is now easy to prove the following formula which turns out to be useful for studying iterates of the quantum propagator [9]: $\forall \ell \in \mathbb{Z}_N$ and $\forall m, n \in \mathbb{Z}$

$$U_N(B)^{-1} T_N(m, n) V_A B \mathbf{e}_\ell = \tag{16}$$

$$\frac{2}{N} \sum_{j \in S_\ell} \sum_{k \in S_{j+n}} e^{-\frac{4i\pi}{N}\ell j} e^{-\frac{i\pi}{N}mn} e^{-\frac{2i\pi}{N}mj} e^{\frac{4i\pi}{N}k(j+n)} \mathbf{e}_k.$$

This follows from a simple calculation using

$$\mathcal{F}_N T_N(m, n) \mathcal{F}_N^{-1} = T_N(n, -m), \tag{17}$$

It is not difficult to recognize in the composition of the two previous Fourier terms the transposition at the quantum level of the cutting-stretching-folding action of B. This can be seen very clearly using the corresponding generating function and use it to define, as usual, the phase of the quantum propagator. We refer to [37] for more about this. Here for simplicity we omit the discussion about which Hilbert space, i.e. which phase (θ_1, θ_2) to chose and set (θ_1, θ_2) = (0,0), even if sometimes anti-periodic conditions (θ_1, θ_2) = (1/2, 1/2) are natural and useful. This last choice is dictated by the two symmetries of the systems as briefly discussed in Example 2 and fully described in [37] (and also [11]). This choice, by the way, is unimportant if one is interested in semiclassical asymptotic. In Fig. 10 we show the action of the quantum baker's map.

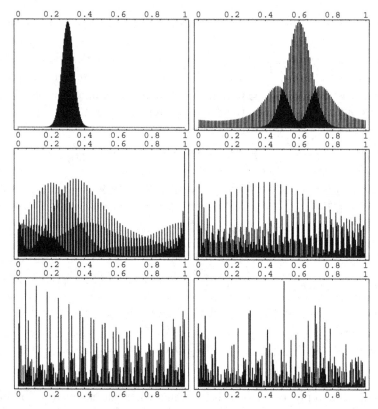

Fig. 10. Evolution of a given initial state ψ (*top left*) under the action of the quantum baker's map $U_N(B)^t$. $N = 244$ and B as in Example 2. $t = 0, 1, 2, 3, 4, 30$ from *top left to bottom right*.

The Sawtooth Map

Given $a, b \in \mathbb{R}$, we consider the following discontinuous maps (for $a, b \in \mathbb{R} \setminus \mathbb{Z}$) A_1, A_2 on the torus:

$$A_1 : (q, p) \longrightarrow (q + ap, p),$$

$$A_2 : (q, p) \longrightarrow (q, p + bq).$$

and we set $A = A(a, b) = A_1 \circ A_2$.

For the classical properties of this maps, we refer to [7, 26] and [38]. As showed in [11], we have the corresponding quantum operators V_1, V_2 in the position representation:

$$V_1 = \mathcal{F}_N^{-1} \circ D_1 \circ \mathcal{F}_N, \qquad V_2 = D_2,$$

where

$$D_1 = \begin{pmatrix} e^{-i\pi N a p_0^2} & \cdots & 0 \\ 0 & \ddots & 0 \\ 0 & \cdots & e^{-i\pi N a p_{N-1}^2} \end{pmatrix},$$

and,

$$D_2 = \begin{pmatrix} e^{i\pi N b q_0^2} & \cdots & 0 \\ 0 & \ddots & 0 \\ 0 & \cdots & e^{i\pi N b q_{N-1}^2} \end{pmatrix}.$$

Finally, we set

$$V_A = V_1 \circ V_2.$$

3 Equidistribution of Eigenfunctions

Assume here that a given symplectic dynamics F of $\mathbb{T}^2$ is given, together with the corresponding family (in N) of quantum operators $U_N(F) = U_F$. We denote by $\sigma_N(U_F) = \{\lambda_1, \ldots, \lambda_m\} \subset S^1$ the corresponding spectrum, where λ_j is an eigenvalue with multiplicity d_j (note that $\sum_{j=1}^m d_j = N$).

Avoiding for now the discussion about the distribution of the degeneracies, we assume that an orthonormal set of eigenfunctions $\{\phi_1^{(N)}, \ldots, \phi_N^{(N)}\}$ is given.

From a semiclassical point of view, one of the main problem is to describe the possible limits of the expected values of the observable when $N \to \infty$.

More precisely, we say that a probability measure μ of $\mathbb{T}^2$ is recovered in the classical limit if and only if there exists a sequence of eigenfunctions $\{\phi_{j_k}^{(N_k)}\}_{k \geq 1}$, $N_{k+1} > N_k$ such that for each $f \in C^\infty(\mathbb{T}^2)$

$$\lim_{k \to \infty} \langle \phi_{j_k}^{(N_k)}, Op_\hbar(f) \phi_{j_k}^{(N_k)} \rangle = \int_{\mathbb{T}^2} f \, d\mu.$$

A main subject in the theory of quantum chaos is to understand the structure of the space of such measures, and its relations with the (convex and non empty) set of all invariant measures for F.

We recall that for a genuine symplectic Anosov map, the set of invariant maps does have a rich structure and it does contain far more invariant ergodic measures than the Lebesgue measure m and the atomic ones concentrated on periodic orbits.

In order to explore this problem, note first of all that it is possible to see that the "Egorov property" implies that any weak limit of matrix elements must be an invariant measure for the classical map; the Egorov theorem states that for any suitable nice observable f, say $f \in C^\infty(\mathbb{T}^2)$:[4]

---

[4] For discontinous maps the uniform strong convergence is in general not true and we should rely on a weaker condition, which correspond (roughly speaking) in assuming convergence to zero of the operator $E_F(f)$ on asymptotically large (density 1) subspaces of $\mathcal{H}_N$.

$$\| E_F(f) \| := \| U_F^{-1} Op_\hbar(f) U_F - Op_\hbar(f \circ F) \| \to_{N \to \infty} 0.$$

Here are two basic questions

(i) Which invariant classical measures can we reproduce in the classical limit?

(ii) In particular, if μ is an atomic invariant measure concentrated on periodic orbits, can we reconstruct it from the eigenfunctions in the classical limit (*scars*)?

This problem is appealing from the mathematical point of view, via its relations to problems and conjectures in number theory and in the theory of L-functions as for the "arithmetic" case of hyperbolic automorphisms. It is also very important from a physical point of view, being intimately related to the conceptual meaning of any semiclassical theory.

As it is well known, this kind of questions does represent a highly non trivial mathematical problem also in the case of the geodesic flow on compact surfaces with constant negative curvature (see [35] for a survey).

It has been conjectured that in presence of strong chaoticity for the classical dynamics, e.g. uniform hyperbolicity and/or strongly mixing, only the natural Lebesgue (or in general Liouville) measure should appear in the classical limit. This is basically the so called *quantum unique ergodicity conjecture*, proposed for the first time in [34] for geodesic flows of surfaces of constant negative curvature. In particular there should be no localization around periodic orbits, i.e. no *scars*.

In the case of the geodesic flow, as for a broad class of toral maps, a first partial answer can be given by using basically only classical ergodicity (with respect to the Lebesgue measure) and some minimal information about eigenfunctions in the average.

In fact, a first result is the *Schnirelman's theorem* which (in this case of the torus) states that the Lebesgue measure is recovered with probability 1, i.e. for *almost all* sequences of eigenfunctions. This result has been proven first for ergodic geodesic flows [8, 39] and also for suitable Schrödinger operators corresponding to classical hamiltonian over $T\mathbb{R}^n$ [16]. Equidistribution of almost all eigenfunctions for linear maps on $\mathbb{T}^2$ and their smooth perturbations has been proven in [2], and in [9] for a class of discontinous maps.

It is important to remark that *Schnirelman* type results are a combinations of quite general properties of the systems: classical *ergodicity*, the *Egorov* property and an average result concerning the eigenfunctions. In particular only ergodicity for the classical flow is required, and one does not need to get into the fine structure of individual eigenstates in order to prove equidistribution with probability one.

Moreover, when the classical map is uniquely ergodic, i.e. it admits only the Lebesgue measure as invariant measure, the *quantum unique ergodicity* is clearly almost trivial. This is for example the case of parabolic linear maps as studied in [29].

Violation of QUE can then happen only on very *exceptional sequences* of eigenfunctions. For the case of linear hyperbolic maps on the torus, one can rely on number theory in order to gain extra information about individual eigenfunctions and try to prove or disprove quantum unique ergodicity (see [10] for a very recent breakthrough in this direction: here the authors do in fact prove the existence of these exceptional sequences).

We consider now $A \in SL(2, \mathbb{Z})$ and we focus on the methods and results concerning equidistribution of eigenfunctions for this linear case.

As we will see more precisely later on, the spectrum of $U_N(A)$ has systematic (*arithmetic*) degeneracies related to the inverse of the period $p(N)$ of $U_N(A)$ and this makes the classical limit difficult. As we will see more precisely in the next paragraph, because the Egorov Theorem is exact, i.e.

$$U_A^{-1} Op_\hbar(f) U_A = Op_\hbar(f \circ A), \quad \forall f \in C^\infty(\mathbb{T}^2),$$

$p(N)$ turns out to be naturally identified with $\text{ord}(A, N)$, the *order* or period of A modulo N, that is the least integer $k \geq 1$ for which $A^k = I$ mod N .[5]

In [13] it has been proved that if instead of looking at all integers N, one restrict the classical limit to a *sparse* sequence of primes for which the degeneracies are bounded and they satisfy an additional algebraic properties (*splitting*[6]) in the quadratic extension of the rationals containing the eigenvalues of A, than the only possible limit for the matrix elements of *all* eigenfunctions must coincide with the Lebesque measure.

In this case one can directly relate any single eigenfunctions with some exponential sums over the finite field $\mathbb{Z}_N$, the so called *Kloosterman sums* and then use the Riemann Hypothesis for a curve over a finite field.

It is important to remark that as a consequence of the Riemann Hypothesis, the set of primes for which the degeneracies are at most M and which are split primes for A has positive density which depend on M (see also the discussion on Artin's conjecture in Rudnick's contribution).

These results has been improved in [23], where the following has been proven:

Theorem 3 ([23]). *Let $A \in SL(2, \mathbb{Z})$ an arbitrary hyperbolic automorphism ($|TrA| > 2$). There is a set $\mathcal{N} \subset \mathbb{N}$ of density one so that all eigenfunctions of $U_N(A)$ are equidistributed as $N \to \infty$, $N \in \mathcal{N}$.*

We point out the fact that a large order of A mod N corresponds to classical periodic orbits lying on invariant lattices with large period. Classical hyperbolicity implies equidistribution of "almost all" of these orbits with respect the Lebesgue measure (see Knauf's contribution).

A basic ingredient in the proof of the Theorems of [13] and [23] is that for the hyperbolic automorphisms we will consider: $U_N(A^k) = U_N(A)^k$ (see (3)).

---

[5] Essentially because of the projective nature of the representation $A \to U_N(A)$, we should sometimes consider $\text{ord}(A, 2N)$ instead, but we will try not to be to pedantic on this.

[6] For a fixed $A \in SL(2, \mathbb{Z})$ hyperbolic, the *splitting* conditions is verified for half of the primes

In particular, powers of A generate a family of unitary map which commute with the original quantum propagator $U_N(A)$.

This fact has been extended in [24] where it has been shown, as we will discuss, that there exists (for each fixed N) a whole commutative group of unitary operators, the so called *Hecke operators*, which commute with $U_N(A)$. As a generalization of the previous Theorem, one can then show that joint eigenfunctions of $U_N(A)$ and of these Hecke operators, called *Hecke eigenfunctions* in analogy with the setting of arithmetic modular surfaces [34], are all equidistributed in the classical limit.

Before discussing the proofs of the equidistribution properties of eigenfunctions, we will now discuss the distribution of $\mathrm{ord}(A, N)$ (or $\mathrm{ord}(A, 2N)$) and the construction of the Hecke operators.

3.1 The Period of A mod N and the Hecke Operators

For a given $A \in \Gamma_\theta(2N)$, there exists an integer p such that $A^p = \begin{pmatrix} 1 & 0 \\ 0 & 1 \end{pmatrix}$ mod $2N$, i.e. p is the common period for all points in phase space with rational coordinates of denominator $2N$. This implies:

$$U_N(A)^p = e^{2\pi i \sigma} I$$

for some constant phase σ (depending on N).

This restricts the N eigenvalues of $U_N(A)$ to lie on the p possible sites:

$$\{\exp[\frac{2\pi i(m + \sigma)}{p}] \mid 0 \leq m \leq p - 1\}.$$

In general $p(N) \neq N$, that is, there is no one-to-one correspondence between eigenvalues and sites. Typically, for a hyperbolic map ($|\mathrm{Tr}A| > 2$), there are both unoccupied and multiply occupied sites and this distribution follows the highly irregular behavior of $p = p(N)$ as a function of N (see Fig. 5).

In general, let $A \in SL(2, \mathbb{Z})$ be a hyperbolic matrix ($|\mathrm{Tr}A| > 2$). As it is well known and easy to prove, all periodic orbits do coincide with the points on the torus with rational coordinates. In particular, any discrete lattice $L_N \subset \mathbb{T}^2$ given by

$$L_N = \left\{ (\frac{j}{N}, \frac{k}{N}), \; j, k \in \mathbb{Z}_N \right\}$$

is in fact invariant. We want to understand the period of the map restricted to these invariant lattices and we pose the following:

Definition 2. *We denote* $\mathrm{ord}(A, N)$ *the order (or period) of the map A modulo N. Namely,* $\mathrm{ord}(A, N)$ *is the smallest positive integer k, such that* $A^k = I \bmod N$

The following hold:

$$\text{ord}(A, MN) = \text{lcm}\left(\text{ord}(A, N), \text{ord}(A, M)\right) \quad \text{for any coprime } M \text{ and } N$$

In particular, if $N = \prod_k p_k^{n_k}$:

$$\text{ord}(A, N) = \text{lcm}\left(\text{ord}(A, p_1^{n_1}), \ldots, \text{ord}(A, p_k^{n_k}), \ldots\right)$$

Remark 2. For what concern the semiclassical properties of eigenfunctions and the asymptotic behavior of the energy levels degeneracy, we anticipate the following facts that we will see in more detail later:

(i) On any *prime* lattice (i.e. a lattice with prime denominator), each periodic orbits (besides the fixed point at the origin) share the *same* period. As we will discuss (see also [5,18,32], if N is prime then $\text{ord}(A, N) = (N \pm 1)/m$, where m is a positive integer which coincide basically with the quantum degeneracy [13]. This observation is important in proving equidistribution of eigenfunctions along *prime sequences*.

(ii) As it will appear clearly, the order of $A \in SL(2, \mathbb{Z})$ modulo N is in fact closely related to the problem of finding the order of a given element $x \in \mathbb{Z}_N$.

Let us discuss briefly this problem, referring the reader to any classical text on number theory for more on this: given a prime p and an element $x \in \mathbb{Z}_p$, we denote by $\text{ord}_p(x) \leq p - 1$ the multiplicative order of x in $\mathbb{Z}_p$. The *primitive* elements of $\mathbb{Z}_p$ are the ones for which $\text{ord}_p(x) = p - 1$, i.e. they are generator for the multiplicative group $\mathbb{Z}_p^*$.

Conjecture 1. (Artin) Let $b \in \mathbb{Z}$ such that $b \neq x^2$. Then b is a primitive root (i.e. it has the largest period) for a positive proportion of the primes, i.e.

$$\frac{\#\{p \in \mathcal{P}, \, p \leq n \, : \, \text{ord}_p(b) = p - 1\}}{\pi(n)} \to_{n \to \infty} c(b) > 0$$

where $\mathcal{P}$ is the set of prime numbers, $\pi(n)$ the number of primes smaller than n and the constant $c(b)$ is a b-dependent Euler product.

If we want results which hold for a full proportion of primes then we have to rely on theorems which probably keep us away from the "optimal" lower bound for the generic period [14]:

Theorem 4. *If $b \neq 0, \pm 1$ then there exists $\delta > 0$ such that the following holds for a density 1 set of primes p*

$$\text{ord}_p(b) \geq p^{1/2} \exp((\log p)^{\delta})$$

If we *assume* the Generalized Riemann Hypothesis (GRH), Artin's conjecture holds true and we can strengthen the previous result [14]

Theorem 5. *If GRH is true and* $f : \mathbb{R}^+ \to \mathbb{R}^+$ *is any function diverging to* ∞, $\mathrm{ord}_p(b) > p/f(p)$ *with probability* 1, *i.e. for a full density set of primes*

GRH allows to prove a lower bound for most integers,

Theorem 6. *[22] Let* $b \neq 0, \pm 1$ *be an integer. Assuming GRH, the set of integers* N *such that* $\mathrm{ord}_N(b) \gg N^{1-\epsilon}$ *has order one*

Going back to our maps over the torus, the generalization of this kind of results to the order of hyperbolic matrices [7] $A \in SL(2, \mathbb{Z})$ modulo N does have, as we already stressed, useful consequences in the study of semiclassical properties of toral automorphisms.

On the one hand, lower bounds on the period are crucial in the proof of equidistribution of eigenfunctions for a density 1 set of integers $N \approx \hbar^{-1}$. On the other hand, possible exceptions to equidistribution (*scars*) for hyperbolic linear maps must be found,if any, among the classical limits performed over sets of integers which lead to *very small* order for a given $A \in SL(2, \mathbb{Z})$.

Let us now see better how $\mathrm{ord}(A, n)$ can be studied by using techniques out of algebraic number theory. Basic references are [32] and [5].

As we already stressed, studying $\mathrm{ord}(A, N)$ is equivalent to classify the orbits of A on the lattice $L_N \cong \mathbb{Z}_N^2$. We will show now how, using ideal theory, this lattice can be provide with a natural multiplication which makes it into a ring.

The most important consequence of this is that any question regarding periodic orbits over L_N can in fact be formulated in terms of *ideals*[8] in these rings.

More precisely, the eigenvalues λ, λ^{-1} of A generate a real quadratic field extension $\mathbb{Q}(\lambda) = \mathbb{Q}(\sqrt{D_A})$, where the *field discriminant* D_A is given by the square-free factor of the discriminant $(\mathrm{Tr}A^2 - 4)$, $|\lambda| > 1$ and $|\mathrm{Tr}A| > 2$.

Adjoining $\sqrt{D_A}$ to $\mathbb{Z}$ gives an *order* $\mathcal{D} = \mathbb{Z}[\sqrt{D_A}]$, contained in the ring of integers of $\mathbb{Q}(\sqrt{D_A})$. In general, the ring of integers does not coincide with $\mathbb{Z}[\sqrt{D_A}]$ but it is general larger, depending on $D_A \bmod 4$. In order to avoid some technical complications, we will now always assume that $\mathbb{Z}[\lambda]$ equals $\mathbb{Z}[\sqrt{D_A}]$ and that they do coincide with the full ring of integers of $\mathbb{Q}(\lambda)$.

The elements of $\mathcal{D}$ are called *quadratic integers*, they are given by the numbers of the form $x + y\sqrt{D_A}$, $x, y \in \mathbb{Z}$ and "naturally" define a two-dimensional lattice: $x + y\sqrt{D_A} \to (x, y) \in \mathbb{Z}^2$.

In analogy with complex number, given a quadratic integer $z = x + y\sqrt{D_A}$, we denote by $\bar{z} = x - y\sqrt{D_A}$ its *conjugate* and by $\mathcal{N}(z) = z\bar{z}$ its *norm*. Note that $\mathcal{N}(\bar{z}) = \mathcal{N}(z)$ and $\mathcal{N}(zu) = \mathcal{N}(z)\mathcal{N}(u)$. One can now verify that λ is a

---

[7] If A is elliptic ($|\mathrm{Tr}(A)| < 1$) then A has a finite order not greater than 6. If A is parabolic ($|\mathrm{Tr}(A)| = 1$), then $\mathrm{ord}_N(A) > c_A N$ for a given constant $c_A > 0$

[8] An ideal is a subset of a ring which is closed under addition and which is also invariant under multiplication by any element of the ring

unit in $\mathcal{D}$, i.e. $\mathcal{N}(\lambda) = 1$. Moreover, a non-unit quadratic integer z is called a *prime* if and only if $z = uv$ implies that either u or v is a unit.

$\mathcal{D}$ belong to a class of rings called *Dedekind domain* for which in general *unique factorization into primes fail* (see next example). In order to restore unique factorization one has to introduce the concept of *ideals*, and Dedekind was in fact the first to recognize this (see [5, 32] and references therein).

Example 4. Here we just recall some canonical examples of the previous statements. For instance $z = 2 + \sqrt{7}$ is a prime and the decomposition $2 + \sqrt{7} = (8 + 3\sqrt{7})(-5 + 2\sqrt{7})$ is not a contradiction because $\mathcal{N}(+8 + 3\sqrt{7}) = 1$.

Finally, as an example of the failure of unique prime factorization consider in $\mathbb{Z}[\sqrt{-5}]$

$$21 = 3 \cdot 7 = (1 + 2\sqrt{-5})(1 - 2\sqrt{-5}).$$

It is easy to see that all four factors are primes in $\mathbb{Z}[\sqrt{-5}]$.

Given now a positive integer N, we will denote by (N) the corresponding ideal in $\mathcal{D}$, namely $(N) = \{N(x + y\sqrt{D_A}), \ x, y \in \mathbb{Z}^2\}$.

Using now the map $(x, y) \to x + y\sqrt{D_A}$ with x and y taken mod N, we can then establish the identification:

$$L_N \equiv \mathcal{D}/(N).$$

The most important consequences of the previous construction are the following:

(i) As one can easily verify, under this identification, the action of the map A over L_N is replaced by multiplication by λ in $\mathcal{D}/(N)$.

(ii) In particular, if $z_n = x_n + y_n\sqrt{D_A}$ and $z_{n+1} = x_{n+1} + y_{n+1}\sqrt{D_A}$ corresponds to two successive points of an orbit on L_N, then:

$$z_{n+1} \cong \lambda z_n \text{ mod } (N), \quad \text{i.e. } z_{n+1} - \lambda z_n \in (N).$$

(iii) The period p of a given point $(x, y) \in L_N$ is given by the congruence:

$$\lambda^p z \cong z \text{ mod } (N) \qquad z = x + y\sqrt{D_A}.$$

In particular, "dividing" both sides of the equation by z, we get

$$\lambda^p \cong 1 \text{ mod } (N)/(N, z),$$

where $(N)/(N, z)$ is the new ideal obtained by dividing (N) by the ideal (N, z) is the greatest common divisor of (N) and (z), i.e. the smallest ideal containing both N and z.

It is clear now that the classification of the periodic orbits over L_N is related to the possible values of $(N)/(N, z)$, namely it is related to the *ideal factorization* of (N) in $\mathcal{D}$, factorization which is now *unique (!)*. As usual, the theory is built now from the case N prime, where we have in fact a quite clear picture as described by the following proposition. For a proof and further discussion see [5, 13, 32].

Proposition 2. *Assume that D_A, the square-free factor of $((TrA)^2 - 4)$, is given and that $N > D_A$ is any prime number, then*

1. *If $\left(\frac{D_A}{N}\right) = 1$, we say that N splits and we have the ideal decomposition $(N) = (P1)(P2)$. $\operatorname{ord}(A, N)$ divides $N - 1$ and moreover this is also the least period of each point on $L_N \setminus (0, 0)$.*
   If $\operatorname{ord}(A, N) = (N-1)/m$, then there are $2m$ orbits which belongs entirely either to the ideal factor $(P1)$ or to $(P2)$.
   These are called ideal orbits. The remaining $m(N - 1)$ are called free orbits.
2. *If $\left(\frac{D_A}{N}\right) = -1$, we say that N is non splitting or inert, (N) is a prime ideal and $\operatorname{ord}(A, N)$ divides $N + 1$.*
   Also in this case, all points on $L_N \setminus (0, 0)$ share the same period and if $\operatorname{ord}(A, N) = (N + 1)/m$, then there are $m(N - 1)$ free orbits

Remark 3. The *splitting* case as a quite clear interpretation, which is evident even without using quadratic extensions. Roughly speaking, in this case we can take the square root of D_A over Z_N and this means that A can be diagonalized over $Z_N \times Z_N$ (as vector space over the field $\mathbb{Z}_N$)with eigenvalues λ, λ^{-1} lying in $\mathbb{Z}_N$, namely its characteristic polynomial factorizes over $\mathbb{Z}_N$.

Moreover, $\operatorname{ord}(A, N)$ is just the order of $\lambda \in \mathbb{Z}_N$ and the fact that it does divide $N - 1$ is just the little Fermat Theorem (see Rudnick's contribution).

Example 5. We consider again the map $A = \begin{pmatrix} 2 & 1 \\ 3 & 2 \end{pmatrix}$ as in Example 1.

In Fig. 5, we plot $\operatorname{ord}(A, N)$ both for generic integers (left) and also the restriction to prime values of N (right).

For this map, the discriminant field is $D_A = 3$. As one can verify, for example $N = 23, 37, 61, 167, 229, 431, 541, \dots$ are some splitting primes, whereas $N = 29, 31, 67, 89, 137, 223, 353, 439, 523, \dots$ gives some non-splitting lattices. For example, if $N = 229$ then $3 = 71^2 \bmod 229 = (-71)^2 \bmod 229 = 158^2 \bmod 229$.

The real eigenvectors $v = (-1, \sqrt{3})$ and $w = (1, \sqrt{3})$ that $(Av = \lambda v$ and $Aw = \lambda^{-1}w$, $\lambda = 2 + \sqrt{3})$ give rise to the eigenvectors $v_N = (-1, 71) = (228, 71)$ and $w_N = (1, 71)$ for the map A acting on $\mathbb{Z}_N^2$. In particular the orbit of any point z in the $\mathbb{Z}_{229}$-linear subspace $\mathbb{Z}_{229} \cdot v_N \subset \mathbb{Z}_{229} \times \mathbb{Z}_{229}$ is given by $z, \lambda z, \lambda^2 z, \dots$, where now $\lambda = 2 + 71 \in \mathbb{Z}_{229}$ and they coincide with the ideals orbits The same is true for points in $\mathbb{Z}_{229} \cdot w_N \subset \mathbb{Z}_{229} \times \mathbb{Z}_{229}$, where now the action is given by the multiplication of $\lambda^{-1} = 2 - 71 = 160 \in \mathbb{Z}_{229}$. The period in this case turns out to be exactly $p = 228 = N - 1$. See Fig. 12 for an example of *ideal* and *free* orbits over L_{229}. As we see, the *lattice* structure of the ideal orbits is quite evident.

Finally, in Fig. 13 we show the action of the map starting form a given initial distributions over L_{229}. While the action appears chaotic, periodicity reconstructs the initial state after 228 iterations.

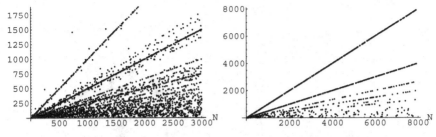

Fig. 11. *Left*: ord(A, N) versus integers $5 \leq N < 3000$. *Right*: ord(A, N) versus N, when N runs over the first 1000 primes.

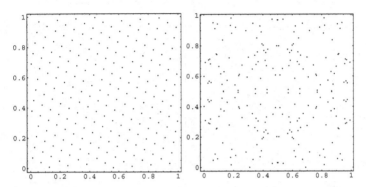

Fig. 12. *Left*: periodic ideal orbit over L_{229} of period $p = 228$, generated by the initial point $z = (228/229, 71/229)$. *Right*: periodic orbit over L_{229} of period $p = 228$, with initial point $z = (228/229, 71/229)$.

Remark 4. The splitting case is just the one considered in [13], where the existence of the integer valued eigenvectors for the action of A mod N allows to construct explicitly a set of eigenvectors of $U_N(A)$, using linear combinations of functions constants on certain linear subspaces of $\mathbb{Z}_N \times \mathbb{Z}_N$ (*ideal lines*, see also [5, 32]). The asymptotic of the matrix elements of observables with respect this basis of eigenfunctions turns out to be directly related to the behavior of *Kloosterman* sums of the form:

$$K(a, b) = \sum_{k \in \mathbb{Z}_N} e^{\frac{2\pi i}{N}(ak + bk^{-1})}$$

or more general exponential sums K, where the exponent is some rational function of $k \in \mathbb{Z}_N$. In both cases the Riemann hypothesis over finite fields gives us uniform bounds of the form $|K| \approx \sqrt{N}$, which turn out to be sufficient for proving equidistribution in the limit $N \to \infty$ for sequences of splitting primes such that $N/\mathrm{ord}(A, N)$ remains bounded (see [13] for further details).

For general integer values of N and generic values of the degeneracies of $U_N(A)$, one has to rely on some more sophisticated techniques to gain information about the asymptotics of the matrix elements.

Fig. 13. Action of the map over L_{229} for a given initial condition. From *top left to bottom right*: $t = 0, 1, 50, 100, 226, 228$.

To be a little bit more precise, both the norm function $\mathcal{N}$ and the embedding $\imath$ map behaves properly modulo (N) and they induce a pair of well defined maps under the quotient by the ideal (N):

$$\imath_N : \mathcal{D}/(N) \to \mathbb{Z}_N, \quad \mathcal{N} : \mathcal{D}/(N) \to GL_2(\mathbb{Z}_N)$$

Matrices in $SL_2(\mathbb{Z}_N)$ which correspond to unit elements give by construction a set $H_N(A)$ of quantum operators which commute with the original $U_N(A)$. These are called *Hecke operators* and, as it is discussed in [24], they show how in general degeneracies for these "arithmetic" quantum systems are related to the existence of quantum symmetries.

Referring the reader to [24] for more precise statements and proofs, we summarize what we need in the following:

Proposition 3. *Given $A \in \Gamma_\theta(2N)$ and for any fixed integer N, let*

$$- \text{ Hecke operators } -$$

$$H_N(A) := \{B \in SL(2, \mathbb{Z}_N) \mid B = \imath_N(\gamma), \, \mathcal{N}(\gamma) = 1\}$$

then:

(i) $\forall B \in H_N(A)$

$$U_N(A) \, U_N(B) = U_N(AB) = U_N(BA) = U_N(B) \, U_N(A).$$

As a consequence, the eigenspace of $U_N(A)$ breaks up into joint eigenspace of the Hecke operators. More precisely, there exists an orthonormal basis $\{\psi_j\}$ of $\mathcal{H}_N$ (Hecke eigenfunctions) and a set of characters α_j of $\imath_N^{-1}(H_N(A))$,for $j = 1, \ldots, N$ such that:

$$U_N(B) \, \psi_j = \alpha_j(\gamma) \, \psi_j, \quad \imath_N(\gamma) = B.$$

(ii) $\forall \epsilon > 0$:

$$N^{1-\epsilon} \ll |H_N(A)| \ll N^{1+\epsilon}$$

By reducing to a counting problem, one can then prove equidistribution of the Hecke eigenfunctions.

Theorem 7. *Let $A \in \Gamma_\theta(2N)$ be a hyperbolic matrix.*

(i) Then for unit norm Hecke eigenfunctions Ψ and $\forall \epsilon > 0$:

$$\langle \psi, Op_N(f) \, \psi \rangle = \int_{\mathbb{T}^2} f(x) \, dm(x) + O_{f,\epsilon}(N^{-1/4+\epsilon}), \quad \text{as } N \to \infty$$

(ii) Let $\{\psi_j\}$ be a basis of Hecke eigenfunctions and $(0,0) \neq n \in \mathbb{Z}^2$:

$$\sum_{j=1}^{N} |\langle \psi_j, T_N(n)\psi_j \rangle|^4 \leq |n|^{16} \, N^{-1+\epsilon}, \quad N \to \infty, \, \forall \epsilon > 0$$

(iii) If $\{\psi_j\}$ is a basis of Hecke eigenfunctions and $n = (n_1, n_2) \in \mathbb{Z}_N \times \mathbb{Z}_N$. Let $I \subset \mathbb{Z}[\lambda]$ be the ideal previously defined (see also [23]) on which multiplication by λ is equivalent to the action of A over $\mathbb{Z}^2$. Namely $I = \mathbb{Z}[v_1, v_2]$, where $v = (v_1, v_2) \in \mathbb{Z} \times \mathbb{Z}$ is the eigenvector corresponding to λ: $vA = \lambda v$.

Finally let $\nu \in \mathcal{D}$ such that $\imath(\nu) = n$ and denote by $Q = Q(A, N, n)$ the set of solutions:

$$Q = \{\beta_j \in \imath^{-1}(H_N(A)) \, : \, \nu(\beta_1 - \beta_2 + \beta_3 - \beta_4) = 0 \ mod \ NI \ \},$$

then:

$$\sum_{j=1}^{N} |\langle \psi_j, T_N(n)\psi_j \rangle|^4 \leq \frac{N}{|H_N(A)|^4} \, \sharp Q$$

(iv) The number of elements of Q, namely the number of solutions of

$$\nu(\beta_1 - \beta_2 + \beta_3 - \beta_4) = 0 \ mod \ NI \ , \quad \beta_j \in \imath^{-1}(H_N(A))$$

is bounded by $O(|\mathcal{N}(\nu)|^8 N^{2+\epsilon})$.

Proof. Let us now discuss now some steps of the proof. First of all, (i) follows from (ii) by using a simple argument using the rapid decay of the Fourier coefficients as shown in Theorem 10 in [24]. Now note that statement (ii) of Proposition 3 gives a lower bound for the elements in $\imath^{-1}(H_N(A)) \gg N^{1-\epsilon}$. This, together with the upper bound on the cardinality of the set of solutions Q given by (iv) allows to see immediately that (ii) follows from (iii).

(iii) is now proven by averaging over the Hecke operators and performing a direct calculations using the commutativity of the Hecke operators. More precisely, let

$$D(n) = \frac{1}{|H_N(A)|} \sum_{B \in H_N(A)} U_N(B)^{-1} T_N(n) U_N(B)$$

It is now easy to calculate the matrix elements of D with respect any orthonormal basis of eigenfunctions:

$$
\begin{aligned}
D_{ij} = \langle \psi_j, D(n)\psi_i \rangle &= \frac{1}{|H_N(A)|} \sum_{B \in H_N(A)} \langle \psi_j, T_N(n \cdot B)\psi_i \rangle \\
&= \frac{1}{|H_N(A)|} \sum_{B \in H_N(A)} \langle \psi_j, U_N(B)^{-1} T_N(n) U_N(B) \, \psi_i \rangle \\
&= \frac{1}{|H_N(A)|} \sum_{B \in H_N(A)} \bar{\alpha}_i(B) \alpha_j(B) \, t_{ij} \\
&= \begin{cases} t_{ij} \text{ if } \alpha_i = \alpha_j \\ 0 \quad \text{otherwise} \end{cases}
\end{aligned}
\tag{1}
$$

Where $t_{ij} = \langle \psi_i, T_N(n)\psi_j \rangle$, and where the last equality follows because the α_j's are non trivial characters.

Now note that:

$$\sum_{\lambda_j = \lambda_i} |t_{ij}|^4 \leq \mathrm{Tr}\left((D^*D)^2 \right).$$

In fact,

$$\text{Tr}(D^*D)^2 = \sum_k \sum_\ell \bar{D}_{\ell k} D_{\ell k} = \sum_k \left(\sum_{\lambda_\ell = \lambda_k} |t_{\ell k}|^2 \right) \geq \sum_{\lambda_j = \lambda_i} |t_{ij}|^4.$$

Now, using $U_N(B)^{-1} T_N(n) U_N(B) = T_N(n \cdot B)$ and $T_N^*(m) = T_N(-m)$, we can substitute the definition of $D(n)$ and see immediately that $(D^*D)^2$ is given by $1/|H_N(A)|^4$ times a sum ranging over all $B_1, B_2, B_3, B_4 \in H_N(A)$, of terms:

$$T_N(nB_1)T_N(-nB_2)T_N(nB_3)T_N(-nB_4) =$$
$$= \gamma(B_1, B_2, B_3, B_4)T_N(n(B_1 - B_2 + B_3 - B_4)), \tag{2}$$

where $|\gamma| = 1$. If now we take the trace and use $\text{Tr}T_N(n) = N \delta_n^{(0,0)}$, we see immediately that

$$\text{Tr}(D^*D)^2 \leq \frac{N}{|H_N(A)|^4} \sharp Q$$

and (iii) follows.

Finally, the counting problem (iv) is a non trivial one but it is now reduced to counting number of solutions of certain equations, similar (but a little bit more complicated) than the one discussed in Rudnick's contribution. A complete proof of this can be found in [24]

We end this section now by recalling some results presented in [24] and [23] concerning upper and lower bound for $\text{ord}(A, N)$.

Theorem 8 ([22]). *Let $A \in SL(2, \mathbb{Z})$ be hyperbolic. Assuming GRH, the set of integers N such that $\text{ord}_N(A) \gg N^{1-\epsilon}$ has density one.*

As far as it concerns the short period lattice, we have the following

Theorem 9 ([23]). *There is an infinite sequence of integers $\{N_k\}_{k=1}^{\infty}$ for which*

$$\text{ord}(A, N_k) \ll \log N_k$$

3.2 Values Distribution for Eigenfunctions

We review here some recent results regarding the equidistribution and values distribution for eigenfunctions of quantized maps.

Again with the use of the Hecke operators associated to U_A, the following result could be proved [25]:

Theorem 10. *Let ψ a normalized Hecke eigenfunction, then*

$$\| \psi \|_\infty \ll_\epsilon N^{3/8+\epsilon}$$

Remark 5. Note that as a consequence of the L^2-normalization

$$\| \psi \|_2^2 = \frac{1}{N} \sum_{q \in \mathbb{Z}_N} |\psi(q)|^2 = 1,$$

we have the trivial bound

$$\| \psi \|_\infty \le N^{1/2}$$

Given $A \in SL(2, \mathbb{Z})$, we denote now by $\mathcal{P}_A$ the set of all splitting primes N. Namely, as described before, $\mathcal{P}_A$ is the set of density $1/2$ such that A is diagonalizable modulo N. In this case, for any $N \in \mathcal{P}_A$, the set of the Hecke operators is isomorphic to the multiplicative group $\mathbb{Z}_N^*$, and the Hecke eigenfunctions correspond to Dirichlet characters χ modulo N. We denote by $\psi_{\chi,N}$ the corresponding normalized eigenfunction. The following result concerning the suprema of the Hecke eigenfunctions is proved in [23].

Theorem 11. *If $N \in \mathcal{P}_A$ and A is not triangular modulo N, then:*

(i) The Hecke eigenspace corresponding to the trivial character χ_0 is one dimensional. Moreover, there exists an orthonormal basis $\{\psi_{0,N}, \psi_{\infty,N}\} \cup \{\psi_{\chi,N}\}_{\chi \ne 1}$, where χ ranges over all non-trivial characters, such that:

$$|\psi_{\infty,N}| = 1, \qquad |\psi_{0,N}| = \sqrt{1 - 1/N}$$

(ii) For non trivial character χ, the corresponding eigenspace is one-dimensional. Moreover,

$$\| \psi_{\chi,N} \| \le 2\sqrt{1 - 1/N}$$

In order to study the value distribution of eigenfunctions, we denote by μ_{sc} the *semi-circle measure* on $[0, 1]$, obtained as the image of the haar measure on $SU(2)$ through the map $g \to |\mathrm{Tr}(g)|/2$. We have the following (see [25])

Theorem 12. *If $N \in \mathcal{P}_A$ and χ is any non-trivial character*

(i)

$$\frac{1}{N}\#\{t \in \mathbb{Z}_N : |\psi_{\chi,N}|/2 \in I\} \to_{N \in \mathcal{P}_A} \mu_{sc}(I) = \int_I \frac{4}{\pi}\sqrt{1 - u^2}\, du.$$

(ii) For $r \ge 2$, and any choice of distinct non trivial characters $\chi_1, \ldots, \chi_r$, the amplitudes $|\psi_{\chi_1,N}|, \ldots, |\psi_{\chi_r,N}|$ are statistical independent. Namely, for any choice of subintervals $I_1, \ldots, I_r \subset [0, 1]$ we have

$$\frac{1}{N}\#\{t \bmod N : |\psi_{\chi_i,N}(t)| \in I_i, \forall i = 1, \ldots, r\} \to \prod_{i=1}^t \mu_{sc}(I_i),$$

as $N \in \mathcal{P}_A \to \infty$.

In [23] the reader can find the proof of these Theorem and an interesting comparison with Maass forms.

3.3 Equidistribution of Eigenfunctions for Discontinuous Maps

Here we will be interested in reviewing some results concerning the classical limit of quantized, discontinuous, ergodic or mixing symplectic transformations of the two-torus. Most of the material here have been recovered from [9]. In particular we will see, following [9], that combining ideas of [40] and [41] with the approach of [2], one can prove an equipartition result for eigenfunctions, provided one has a suitable version of the semi-classical Egorov Theorem (Definition 3).

As we already discussed for linear automorphisms, the semi-classical Egorov Theorem states that quantization and evolution commute up to terms of order at most $\hbar$. It can take several forms. In semi-classical analysis for systems where the underlying classical dynamics is smooth there are two versions of the result (see [33], Theorem 4.10 and Theorem 4.30). The deepest version states that the exact propagator differs by terms of order $\hbar$ from a suitable Fourier integral operator the phase of which is given in terms of the classical action. This implies an easier version which states that the quantization of an evolved classical observable differs from the quantum evolution of the quantized classical observable by terms of order $\hbar$. For our purposes, only the second version is needed. For billiards, where the flow has singularities, a suitably adapted version of this result was proven in [6, 15]. This is precisely the version used in [41] to obtain the Schnirelman Theorem for ergodic billards.

In [9] it has been shown that Egorov estimates, in the sense of Definition 3 below, hold for the quantized sawtooth and Baker maps and are sufficient to derive an equipartition result.

The sawtooth and baker's maps are prototypical discontinuous uniformly hyperbolic systems. The baker is easily seen and well known to be a Bernoulli system (see e.g. [37] and references therein). The dynamical properties of the sawtooth maps are much harder to derive and have been studied extensively in recent years. They are globally hyperbolic discontinuous systems and have been proven to be exponentially mixing and hence in particular ergodic [7, 26, 27, 38]. Their periodic orbits and various other dynamical properties have also been studied in detail (see [38] for further references).

First, there are very few explicit classical symplectic dynamical systems known to be hyperbolic, mixing, or even simply ergodic. In continuous time, we have the geodesic flows on (constant) negative curvature Riemannian manifolds, and ergodic billard flows. In discrete time, there are the hyperbolic automorphisms of the torus (and their perturbations), as well as the aforementioned discontinuous maps. Note that among those only the geodesic flows and the (smoothly perturbed) toral automorphisms are smooth. This has been a major motivation for analyzing systems with singularities.

It seems to be generally believed (or at least hoped) that the quantized sawtooth and Baker maps display "typical" behaviour of quantized chaotic systems [31, 37].

A second reason for the interest in those maps is that the discrete time variable and the finiteness of the quantum Hilbert spaces associated to the torus constitute a clear advantage for numerical – and to some extent theoretical – studies (see Bäcker 's contribution).

There are three important aspects of these systems that we should keep in mind: the singularities in the maps, their chaotic character, and the fact that the finite dimensionality of the quantum Hilbert spaces also has a drawback since it replaces oscillating integrals by oscillating sums, for which a sufficiently complete and powerful stationary phase method is not available [37].

In particular, as we will describe better later on, the singularities in the classical maps introduce new problems making even the Egorov Theorem a non-trivial matter. It can be seen in fact that both for the sawtooth maps and the Baker transformation, the Egorov theorem does not hold in its usual form.

It seems therefore a good idea to start by establishing the Schnirelman theorem for discontinuous ergodic systems.

As we will recall in detail below, it is possible to associate to some symplectic maps T on the two-torus a corresponding quantum operator $U_N(A)$. This operator acts on a suitable Hilbert space $\mathcal{H}_N$ of dimension $N \in \mathbb{N}$, where N is related to the Planck constant via $2\pi\hbar N = 1$. We will not indicate the N dependence of $U_N(A)$.

Given a generic area preserving map A of the torus and a function $f \in C^\infty(\mathbb{T}^2)$ we are interested in the behaviour of the following operator, $\forall k \geq 1$, and $N \to \infty$:

$$E_A^{(k)}(f) = U_N(A)^{-k} Op_N(f) U_N(A)^k - Op_N(f \circ A^k), \qquad (3)$$

where we have assumed that $f \circ T^k \in C^\infty(\mathbb{T}^2)$.

The operators $E_A^{(k)}(f)$ measure the amount to which quantization and evolution do not commute. As we already know $E_A^{(k)}(f) = 0$, if A is a linear map.

For future purposes, note that a simple induction procedure gives: ($f \circ T^\ell \in C^\infty(\mathbb{T}^2)$, $0 \leq \ell \leq k$),

$$E_A^{(k)}(f) = U_N(A)^{-1} E_A^{(k-1)}(f)U_N(A) + E_A(f \circ A^{(k-1)}). \qquad (4)$$

We now give the following key definition.

Definition 3. *Let A be a map on the torus and $U_N(A)$ its quantization. We say $(A, U_N(A))$ satisfies an Egorov estimate up to time K if the following holds:*

1. *there exists a closed set Σ_K of measure zero so that, if $f \in C^\infty(\mathbb{T}^2)$ is supported away from Σ_K, then $f \circ A^\ell \in C^\infty(\mathbb{T}^2)$, $\forall \ell \leq K$;*

2. *for each family of orthonormal bases $\{\psi_j^{(N)}\}_{j=1,\dots,N}$, there exists a family of index sets $\mathcal{E}^{(K)}(N) \subset \{1, 2, \dots, N\}$ satisfying $\mathcal{E}^{(K)}(N)/N \to 1$ so that $\forall 0 \leq \ell \leq K$*

$$\sup_{j \in \mathcal{E}^{(K)}(N)} \| E_A^{(\ell)}(f) \psi_j^{(N)} \| \overset{N \to \infty}{\longrightarrow} 0. \tag{5}$$

The set Σ_K should be thought of as the union of the set of singularities for T with its image under $A, \ldots, A^K$. As $K \to \infty$, it tends to "fill" the torus, in the sense that it cuts the torus into disconnected pieces of increasingly small area, becoming eventually smaller than the elementary area $2\pi\hbar$. The first condition in the definition states that the classical observable must stay away from those. Since we need control for arbitrarily large K, this might look worrisome, since it seems to impose untenable restrictions on f. Fortunately, in the proof of the Schnirelman Theorem, one always takes $\hbar \to 0$ before taking $K \to \infty$, avoiding this difficulty.

Here is the main result in [9]:

Theorem 13. *Suppose A is ergodic and $(A, U_N(A))$ satisfies an Egorov estimate for all times K. Denote by $\varphi_j^{(N)}$ a normalized basis of eigenvectors of $U_N(A)$ and let $f \in C^\infty(\mathbb{T}^2)$. Then*

$$\frac{1}{N} \sum_{j=1}^{N} | \langle \varphi_j^{(N)}, Op_N(f) \varphi_j^{(N)} \rangle - \int f \, dq \, dp |^2 \to 0, \tag{6}$$

which is equivalent to the following statement.

There exists $\mathcal{E}(N) \subset \{1, \ldots, N\}$ with $\mathcal{E}(N)/N \to 1$ so that $\forall j_N \in \mathcal{E}(N)$, $\forall f \in C^\infty(\mathbb{T}^2)$

$$\langle \varphi_{j_N}^{(N)}, Op_N(f) \varphi_{j_N}^{(N)} \rangle \to \int f \, dq \, dp. \tag{7}$$

Equidistribution for almost all eigenfunctions for sawtooth and baker's map (at least for a class of observables) follows now from Theorem 13 and the next Proposition proved in [9].

Proposition 4.

(i) *The quantized sawtooth maps satisfy an Egorov estimate, so that Theorem 13 holds for them.*

(ii) *If one restrict to functions $f \in C^\infty(\mathbb{T}^2)$, depending on q alone, the quantized baker's map satisfy an Egorov estimate and (7) of the previous Theorem holds for these observables.*

Proof of Theorem 13

Here we review the proof of Theorem 13, following closely the arguments in [9].

For $f \in C^\infty(\mathbb{T}^2)$, we write $\bar{f} = \int_{\mathbb{T}^2} f \, dq \, dp$. We will denote the time-average of f and of $Op_N(f)$ by

$$\{f\}_K = \frac{1}{K} \sum_{\ell=0}^{K-1} (f \circ A^\ell),$$

$$\{Op_N(f)\}_K = \frac{1}{K} \sum_{\ell=0}^{K-1} U_N(A)^{-\ell} Op_N(f) U_N(A)^{\ell},$$

and write

$$Z_N(f) = \frac{1}{N} \sum_{j=1}^{N} \left| \langle \varphi_j^{(N)}, Op_N(f - \bar{f}) \varphi_j^{(N)} \rangle \right|^2, \tag{8}$$

for any $f \in C^\infty(\mathbb{T}^2)$ (see [41]). For any fixed K and ϵ to be chosen later, introduce a smooth characteristic function $\chi_{\epsilon,K}$ of Σ_K, with the property that $m(\text{Supp}\,\chi_{\epsilon,K}) \leq \epsilon$, where m denotes the Lebesgue measure. Then, for any $f \in C^\infty(\mathbb{T}^2)$

$$f = f \chi_{\epsilon,K} + f(1 - \chi_{\epsilon,K}). \tag{9}$$

We shall write $f_{\epsilon,K} \equiv f(1 - \chi_{\epsilon,K})$,

$$\gamma_1 = \int_{\mathbb{T}^2} f_{\epsilon,K}\,dm, \qquad \gamma_2 = \int_{\mathbb{T}^2} f \chi_{\epsilon,K}\,dm.$$

Clearly (see 8)

$$Z_N(f) \leq 2\left[Z_N(f_{\epsilon,K}) + Z_N(f\chi_{\epsilon,K})\right]. \tag{10}$$

We control $Z_N(f_{\epsilon,K})$ first. Note that, since $(A, U_N(A))$ satisfies an Egorov estimate up to time K, $(f_{\epsilon,K} \circ A^\ell) \in C^\infty(\mathbb{T}^2)$ for all $\ell \leq K$ and hence $Op_\hbar(f_{\epsilon,K} \circ A^\ell)$ is well defined. Since the φ_j are eigenfunctions of $U_N(A)$ (we drop the N on the $\varphi_j^{(N)}$),

$$Z_N(f_{\epsilon,K}) = \frac{1}{N} \sum_{j=1}^{N} |\langle \varphi_j, \{Op_N(f_{\epsilon,K} - \gamma_1)\}_K \varphi_j \rangle|^2. \tag{11}$$

Noting that $|\langle \varphi_j, B\varphi_j \rangle|^2 \leq \langle \varphi_j, B^*B\varphi_j \rangle$ for any $B \in \mathcal{L}(\mathcal{H}_N)$, we get

$$Z_N(f_{\epsilon,K}) \leq \frac{1}{N} \sum_{j=1}^{N} \langle \varphi_j, \{Op_N(f_{\epsilon,K} - \gamma_1)\}_K^* \{Op_N(f_{\epsilon,K} - \gamma_1)\}_K \varphi_j \rangle. \tag{12}$$

With an eye towards using the Egorov estimate (5) in Definition 3, we rewrite this as follows:

$$Z_N(f_{\epsilon,K}) \leq \frac{1}{N} \sum_{j=1}^{N} \langle \varphi_j, \{Op_N(f_{\epsilon,K} - \gamma_1)\}_K^* \{Op_N(f_{\epsilon,K} - \gamma_1)\}_K \varphi_j \rangle$$

$$= \frac{1}{N} \sum_{j \notin \mathcal{E}^{(K)}(N)} \langle \varphi_j, \{Op_N(f_{\epsilon,K} - \gamma_1)\}_K^* \{Op_N(uf_{\epsilon,K} - \gamma_1)\}_K \varphi_j \rangle$$

$$+ \frac{1}{N} \sum_{j \in \mathcal{E}^{(K)}(N)} \langle \varphi_j, Op_N(\{f_{\epsilon,K} - \gamma_1\}_K)^* Op_N(\{f_{\epsilon,K} - \gamma_1\}_K) \varphi_j \rangle$$

$$+ \frac{1}{N} \sum_{j \in \mathcal{E}^{(K)}(N)} < \varphi_j, [\{Op_N(f_{\epsilon,K} - \gamma_1)\}_K - Op_N(\{f_{\epsilon,K} - \gamma_1\}_K)]^*$$

$$\{Qf_{\epsilon,K} - \gamma_1\}_K \varphi_j >$$

$$+ \frac{1}{N} \sum_{j \in \mathcal{E}^{(K)}(N)} < \varphi_j, \{Op_N(f_{\epsilon,K} - \gamma_1)\}_K^*$$

$$[\{Op_N(f_{\epsilon,K} - \gamma_1)\}_K - Op_N(\{f_{\epsilon,K} - \gamma_1\}_K)] \varphi_j >.$$

Now we use that $\forall f, g \in C^\infty(\mathbb{T}^2)$

$$\| Op_N(f) Op_N(g) - Op_N(fg) \|_{\mathcal{H}_N} \leq \frac{C(f,g)}{N} \tag{13}$$

to conclude that there exists a positive constant $C_{\epsilon,K}(f)$, such that the following estimate holds

$$Z_N(f_{\epsilon,K}) \leq \frac{N - \#\mathcal{E}^{(K)}(N)}{N} C_{\epsilon,K}(f)$$

$$+ \quad \frac{1}{N} \sum_{j \in \mathcal{E}^{(K)}(N)} \langle \varphi_j, Op_N(|\{f_{\epsilon,K} - \gamma_1\}_K|^2) \varphi_j \rangle + \frac{C_{\epsilon,K}(f)}{N}$$

$$+ \quad C_{\epsilon,K}(f) \frac{\#\mathcal{E}^{(K)}(N)}{N} \sup_{j \in \mathcal{E}^{(K)}(N)} \sup_{k \in \{0,\ldots,K-1\}} \| E_A^{(k)}(f_{\epsilon,K} - \gamma_1) \varphi_j \|_{\mathcal{H}_N}.$$

Using that $\left| \frac{1}{N} \operatorname{Tr} Op_N(g) - \int_{\mathbb{T}^2} g\, dm \right| \leq \frac{C(g)}{N}$, for any $g \in C^\infty(\mathbb{T}^2)$, (14) yields, modulo yet another easily controlled error term,

$$Z_N(f_{\epsilon,K}) \leq \int_{\mathbb{T}^2} |\{f_{\epsilon,K} - \gamma_1\}_K|^2 \, dm$$

$$+ C_{\epsilon,K}(f) [\left(1 - \frac{\#\mathcal{E}^{(K)}(N)}{N}\right) + \frac{1}{N}$$

$$+ \frac{\#\mathcal{E}^{(K)}(N)}{N} \sup_{j \in \mathcal{E}^{(K)}(N)} \sup_{k \in \{0,\ldots,K-1\}} \| E_A^{(k)}(f_{\epsilon,K} - \gamma_1) \varphi_j \|].$$

$$\tag{14}$$

Now, choose $\delta > 0$ fixed. Then, since A is ergodic, we have for all K large enough

$$\int_{\mathbb{T}^2} |\{f\}_K - \bar{f}|^2 \, dm < \frac{\delta}{11}. \tag{15}$$

Note that, for such K, $\lim_{\epsilon \to 0} \{f_{\epsilon,K}\}_K = \{f\}_K$ and $\lim_{\epsilon \to 0} \gamma_1 = \bar{f}$, so that the dominated convergence theorem yields that, for ϵ sufficiently small

$$\int_{\mathbb{T}^2} |\{f_{\epsilon,K}\}_K - \gamma_1|^2 \, dm \le \frac{\delta}{4}. \tag{16}$$

Taking K and ϵ as above, and inserting (16) in (14), the Egorov estimate (5) implies that, for N sufficiently large

$$Z_N(f_{\epsilon,K}) \le \frac{\delta}{2}. \tag{17}$$

It remains to control the term $Z_N(f\chi_{\epsilon,K})$. Note that, by using the previous observation regarding the trace of $Op_N(g)$, we can write

$$\left| \frac{1}{N} \mathrm{Tr} \, Op_N \left(|f\chi_{\epsilon,K} - \gamma_2|^2 \right) - \int_{\mathbb{T}^2} |f\chi_{\epsilon,K} - \gamma_2|^2 \, dm \right| \le \frac{1}{N} C_{\epsilon,K}(f). \tag{18}$$

In particular, given $\epsilon > 0$ and K, we can choose N sufficiently large such that (see also (13))

$$\begin{aligned}
Z_N(f\chi_{\epsilon,K}) &\le \frac{1}{N} \sum_{j=1}^N \langle \varphi_j, \, Op_N(f\chi_{\epsilon,K} - \gamma_2)^* \, Op_N(f\chi_{\epsilon,K} - \gamma_2) \varphi_j \rangle \\
&\le \frac{1}{N} \mathrm{Tr} \, Op_N \left(|f\chi_{\epsilon,K} - \gamma_2|^2 \right) + \frac{C'(\epsilon, K)}{N} \\
&\le \int_{\mathbb{T}^2} |f\chi_{\epsilon,K} - \gamma_2|^2 \, dm + \frac{C''(\epsilon, K)}{N} \le \frac{\delta}{2}.
\end{aligned} \tag{19}$$

The first equation in Theorem 13 now follows from (10), (17), and (19). The rest follows from standard diagonalization and density arguments which we omit [8, 39].

Appendix: The Heisenberg Group

We recall here the basic notions about quantization on $\mathbb{R}^{2n}$ via representations of the Heisenberg group.

Consider $\mathbb{R}^{2n+1}$ with coordinates:

$$(p_1, \ldots, p_n, q_1, \ldots, q_n, t) := (p, q, t) := (x, t).$$

Definition 4.

– *The Heisenberg Lie algebra h_n is the vector space $\mathbb{R}^{2n+1}$ with Lie bracket*

$$[(x,t),(y,s)] = (0,0,\omega(x,y))$$

where $x = (p,q)$, $y = (p',q')$ and ω is the usual symplectic $2-$form, i.e.

$$\omega(x,y) = \sum_{l=0}^{n}(p_l q_l' - q_l p_l')$$

– $\mathbb{H}_n(\mathbb{R})$, *the Heisenberg group, is the simply connected Lie group with Lie algebra h_n*

– $Aut(\mathbb{H}_n(\mathbb{R}))$ *and $Aut(h_n)$ are the automorphism groups of $\mathbb{H}_n(\mathbb{R})$ and h_n (as topological group and Lie algebra, respectively).*

Note that if $P_1, ..., P_n, Q_1, ..., Q_n, T$ is the standard basis for $\mathbb{R}^{2n+1}$, the Lie algebra structure is given by

$$[P_i, P_k] = [Q_i, Q_k] = [P_j, T] = [Q_j, T] = 0 \tag{20}$$
$$[P_j, Q_k] = \delta_j^k \cdot T. \tag{21}$$

In both quantum and classical mechanics, the momentum, position and constant observable span a Lie algebra isomorphic to h_n. h_n is a nilpotent Lie algebra and can be identified with a subgroup of $M_{n+2}(\mathbb{R})$ through the map (w.l.g. $n = 1$):

$$\Psi(p,q,t) := \begin{pmatrix} 0 & p_1 & p_2 & t \\ 0 & 0 & 0 & q_1 \\ 0 & 0 & 0 & q_2 \\ 0 & 0 & 0 & 0 \end{pmatrix}.$$

In fact

$$[\Psi(x,t), \Psi(y,s)] = \Psi([(x,t),(y,s)])$$

where on the l.h.s. we have the usual Lie structure on $M_{n+2}(\mathbb{R})$ given by the commutator.

Moreover, because h_n is two-step nilpotent, we immediately get the group law:

$$\exp \Psi(x,t) \cdot \exp \Psi(y,s) = \exp \Psi(x+y, t+s+\frac{1}{2}\omega(x,y)).$$

If $(x,t) \in \mathbb{R}^{2n+1}$ is identified with the matrix $e^{\Psi(x,t)}$, the Heisenberg group is realized as $\mathbb{R}^{2n+1}$ with the group law

$$(x,t) \cdot (y,s) = (x+y, t+s+\frac{1}{2}\omega(x,y)).$$

With this identification, the exponential map is simply the identity and the inverse element of (x,t) is given by $(-x,-t)$. The center of the group is clearly given by:

$$\mathbb{Z}_n = \{(0,0,t) : t \in \mathbb{R}\}.$$

Let us now recall that the unitary representations of $\mathbb{H}_n(\mathbb{R})$ on $L^2(\mathbb{R}^n, \mu)$, where μ is the Lebesgue measure, are defined as follows:

$$(T_\hbar(p,q,t)f)(x) := e^{2\pi i h t + 2\pi i q x + \pi i h p q} \cdot f(x + hp)$$

$\forall h \neq 0$ and $\forall f \in L^2(\mathbb{R}^n, \mu)$, $\forall (p,q,t) \in h_n$

For any $h \in \mathbb{R}$, $T_\hbar$ is a unitary representation, and $T_\hbar$ is not equivalent to $\widehat{T}_{h'}$ for $h \neq h'$ because the central characters $e^{2\pi i h t}$ and $e^{2\pi i h' t}$ are inequivalent. For any fixed h, $T_\hbar$ is the Schrödinger representation: up to equivalences, it is the only unitary irreducible representations of $\mathbb{H}_n(\mathbb{R})$ non-trivial on the center for $h \neq 0$ (Stone-von Neumann Theorem) [15].

If $h = 0$ the (irreducible) representation factors through the quotient group

$$\mathbb{H}_n(\mathbb{R})/\mathbb{Z}_n \simeq \mathbb{R}^{2n}$$

namely the representation is one-dimensional (Schur's lemma) and hence a homomorphism of $\mathbb{R}^{2n}$ into the circle. Hence we have the following:

Theorem 14. *If π is an irreducible unitary representation of $\mathbb{H}_n(\mathbb{R})$, then it is equivalent to one and only one of the following representations:*

- $T_\hbar$ *acting on* $L^2(\mathbb{R}^n, \mu)$ $h \neq 0$
- $\sigma_{ab}(p,q,t) = e^{2\pi i (ap+qb)}$ $a, b \in \mathbb{R}^n$ *acting on* $\mathbb{C}$

The canonical quantization of any classical observable, i.e. any smooth phase-space function $f : \mathbb{R}^{2n} \to \mathbb{C}$, is then obtained as follows: given any suitably smooth function $f : \mathbb{R}^{2n} \to \mathbb{C}$, written via Fourier representation in the following way

$$f(p,q) = \int_{\mathbb{R}^{2n}} \tilde{f}(\eta, \xi) \cdot e^{2\pi i (\eta q + \xi p)} d\eta d\xi$$

the corresponding element $Op_\hbar(f)$ in the algebra is then defined as

$$Op_\hbar(f) = \int \tilde{f}(\eta, \xi) \cdot T_\hbar(\xi, \eta) d\eta d\xi.$$

k Functions in the Hilbert space, i.e. matrix elements of the quantum observables, are related to functions in the phase space $\mathbb{R}^{2n} \simeq T^* \mathbb{R}^n$, i.e. to classical observables, through the matrix elements of the representation $T_\hbar$, namely through the Wigner function and its Fourier transform [13].

References

1. V.I. Arnold, A. Avez, *Ergodic Problems in Classical Mechanics*, Benjamin, New York (1968)
2. A.Bouzouina, S. De Bièvre, *Equipartition of the eigenfunctions of quantized ergodic maps on the torus*, Commun.Math.Phys. **178**,83-105 (1996).

3. M.V. Berry and J.H. Hannay, *Quantization of linear maps on a torus - Fresnel diffraction by a periodic grating*, Physica D **1** (1980), 267-291.

4. N.L. Balazs, A. Voros, *The Quantized Baker's Transformation*, Annals of Physics **190** (1989), 1-31.

5. M. Bartuccelli and F. Vivaldi, *Ideal orbits of toral automorphisms*, Phys. D **39** (1989), 194-204.

6. J. Chazarain, *Construction de la paramétrix du problème mixte hyperbolique pour l'équation d'ondes*, C.R. Acad. Sci. Paris, **276** (1973), 1213-1215.

7. N. Chernoff, *Ergodic and statistical properties of piecewise linear hyperbolic automorphisms of the two-torus*, J.Stat.Phys. **69** (1992), 111-134.

8. Y. Colin de Verdière , *Ergodicite et fonctions propres du laplacien*, Comm. Math. Physics **102**, (1985), 497-502.

9. S. De Bièvre and M. Degli Esposti , *Egorov theorems and equidistribution of eigenfunctions for quantized sawtooth and Baker maps*, Ann. Inst. Henri Poincaré **69**, (1998), 1-30.

10. S. De Bièvre, F. Faure and F. Nonnenmacher , *Scarred eigenstates for quantum cat maps of minimal periods*, nlin.CD/0207060 (2002).

11. S. De Bièvre, M. Degli Esposti and R. Giachetti,*Quantization of a class of piecewise affine transformations on the torus*, Comm. Math. Phys. **176**, (1995), 73-94.

12. M. Degli Esposti, *Quantization of the orientation preserving automorphisms of the torus*, Ann. Inst. Henri Poincaré **58** (1993), 323-341.

13. M. Degli Esposti, S. Isola and S. Graffi, *Classical limit of the quantized hyperbolic toral automorphisms*, Comm.Math.Phys. **167** (1995), 471-507.

14. P. Erdos and M. R. Murty, *On the order of a (mod p)* , In*Number Theory* , Amer.Math. Soc. (1999), 87-97.

15. M. Farris, *Egorov's theorem on a manifold with diffractive boundary*, Comm. P.D.E. **6**, 6 (1981), 651-687.

16. B.Helffer,A.Martinez,D.Robert, *Ergodicité et limite semi-classique*, Commun. Math. Phys. **131** (1985) 493-520.

17. J. Keating, *The cat maps: quantum mechanics and classical motion*, Nonlinearity **4** (1990), 309-341.

18. J. Keating, *Asymptotics properties of the periodic orbits of the cat maps*, Nonlinearity **4**, 277-307 (1990)

19. J. P. Keating, F. Mezzadri and J. M. Robbins, *Quantum boundary conditions for toral maps*, Nonlinearity **12**, 579-591 (1999)

20. J. P. Keating and F. Mezzadri, *Pseudo-symmetries of Anosov maps and spectral statistics*, Nonlinearity **13**, 747-775 (2000)

21. S. Klimek, A. Losniewski, N. Maitra and R.Rubin *Ergodic properties of quantized toral automoprhims*, J. Math. Phys. **38** (1997), 67-83.

22. P. Kurlberg, *On the order of unimodular matrices modulo integers*, arXiv:math.NT/0202053 (2001).

23. P. Kurlberg and Z. Rudnick, *On Quantum Ergodicty for Linear Maps of the Torus*, Commun. Math. Phys. **222**, 201-227 (2001)

24. P. Kurlberg and Z. Rudnick, *Hecke theory and equidistribution for the quantization of linear maps of the torus*, Duke Mathematical Journal **103** (2000), 47-77.

25. P. Kurlberg and Z. Rudnick, *Value distribution for eigenfunctions of desymmetrized quantum maps*, Int. Math. Res. Not. **18**,985-1002 (2001)

26. C. Liverani, *Decay of correlations*, Annals of Mathematics **142** (1995), 239-301.
27. C. Liverani, M.P. Wojtkowski, *Ergodicity in Hamiltonian systems*, to appear in Dyn. Rep. **4**.
28. A. Manning, *Classification of Anosov Diffeomorphism on Tori*, Lecture Notes in Mathematics, No 468 (Springer-Berlin), (1975).
29. J. Marklof and Z. Rudnick, *Quantum unique ergodicity for parabolic maps*, to appear in Geom. Funct. Analysis (2000)
30. F. Mezzadri, *On the multiplicativity of quantum cat maps*, Nonlinearity **15**, 905-922 (2002)
31. P. O'Connor, R. Heller, S. Tomsovic, *Semi-classical dynamics in the strongly chaotic regime: breaking the log-time barrier*, Physica D **55** (1992), 340-357.
32. I. Percival and F. Vivaldi, *Arithmetical properties of strongly chaotic motions*, Phys. D **25** (1987), 105-130.
33. D. Robert, *Autour de l'approximation semi-classique*, Birkhaüser Boston (1987).
34. Z. Rudnick and P. Sarnak, *The behaviour of eigenstates of arithmetic hyperbolic manifolds*, Comm. Math. Physics **161**(1994) 195-213.
35. P. Sarnak, *Arithmetic quantum chaos*, Tel Aviv Lectures (1993).
36. M. Saraceno, *Classical structures in the quantized Baker transformation*, Annals of Physics **199** (1990), 37-60.
37. M. Saraceno, A. Voros, *Towards a semiclassical theory of the quantum baker's map*, Physica D **79** (1994), 206-268.
38. S. Vaienti, *Ergodic properties of the discontinuous sawtooth map*, J. Stat. Phys. **67**, (1992), 251.
39. S. Zelditch, *Uniform distribution of eigenfunctions on compact hyperbolic surfaces*, Duke Math. J. **55**, (1987) 919-941.
40. S. Zelditch, *Index and dynamics of quantized contact transformations*, Ann. Inst. Fourier (Grenoble) **47**, (1997) 305-363.
41. S. Zelditch, M. Zworski, *Ergodicity of eigenfunctions for ergodic billiards*, Commun. Math. Phys. **175**, (1996) 3, 673-682.

Numerical Aspects of Eigenvalue and Eigenfunction Computations for Chaotic Quantum Systems

Arnd Bäcker

Abteilung Theoretische Physik Universität Ulm, Albert-Einstein-Allee 11, D-89081 Ulm, Germany, *arnd.baecker@physik.uni-ulm.de*

Summary. We give an introduction to some of the numerical aspects in quantum chaos. The classical dynamics of two–dimensional area–preserving maps on the torus is illustrated using the standard map and a perturbed cat map. The quantization of area–preserving maps given by their generating function is discussed and for the computation of the eigenvalues a computer program in Python is presented. We illustrate the eigenvalue distribution for two types of perturbed cat maps, one leading to COE and the other to CUE statistics. For the eigenfunctions of quantum maps we study the distribution of the eigenvectors and compare them with the corresponding random matrix distributions. The Husimi representation allows for a direct comparison of the localization of the eigenstates in phase space with the corresponding classical structures. Examples for a perturbed cat map and the standard map with different parameters are shown.

Billiard systems and the corresponding quantum billiards are another important class of systems (which are also relevant to applications, for example in mesoscopic physics). We provide a detailed exposition of the boundary integral method, which is one important method to determine the eigenvalues and eigenfunctions of the Helmholtz equation. We discuss several methods to determine the eigenvalues from the Fredholm equation and illustrate them for the stadium billiard. The occurrence of spurious solutions is discussed in detail and illustrated for the circular billiard, the stadium billiard, and the annular sector billiard.

We emphasize the role of the normal derivative function to compute the normalization of eigenfunctions, momentum representations or autocorrelation functions in a very efficient and direct way. Some examples for these quantities are given and discussed.

1 Introduction

In this text, which is an expanded version of lectures held at a summer school in Bologna in 2001, we give an introduction to some of the numerical aspects in quantum chaos; some of the sections on the boundary integral method contain more advanced material. In quantum chaos one studies quantum systems whose classical limit is (in some sense) chaotic. In this subject computer experiments play an important role. For integrable systems the eigenvalues and eigenfunctions can be determined either explicitly or as solutions of simple equations. In contrast, for chaotic systems there are no explicit formulae for eigenvalues and eigenfunctions such that numerical methods have to be used. In many cases numerical observations have lead to the formulation of important conjectures. Such numerical computations also allow us to test analytical results which have been derived under certain assumptions or by using approximations.

An important class of systems for the study of classical chaos are area–preserving maps as several types of different dynamical behaviour like integrable motion, mixed dynamics, ergodicity, mixing or Anosov systems can be found. We discuss the numerics for the corresponding quantum maps and illustrate some of the methods and results using the standard map and the perturbed cat map as prominent examples.

Another important class of systems are classical billiards and the corresponding quantum billiards. In Sect. 3 we discuss in detail the boundary integral method, which is one of the main methods for the solution of the Helmholtz equation, which is the time–independent Schrödinger equation for these systems.

2 Area Preserving Maps

2.1 Some Examples

We will restrict ourselves to area-preserving maps on the two-torus

$$P : \mathbb{T}^2 \to \mathbb{T}^2 \tag{1}$$

$$(q, p) \mapsto (q', p') \ , \tag{2}$$

where $\mathbb{T}^2 \simeq \mathbb{R}^2/\mathbb{Z}^2$, i.e. the map is defined on a square with opposite sides identified. The requirement that the map P is area–preserving is equivalent to the condition that $\det DP = 1$, where DP is the linearization of the map P. The natural invariant measure on $\mathbb{T}^2$ is the Lebesgue measure $d\mu = dqdp$.

As a first example let us consider the so-called *standard map*, defined by

$$\begin{pmatrix} q' \\ p' \end{pmatrix} = \begin{pmatrix} q + p - \frac{\kappa}{2\pi} \sin(2\pi q) \\ p - \frac{\kappa}{2\pi} \sin(2\pi q) \end{pmatrix} \bmod 1 \ . \tag{3}$$

One easily checks that this map is area-preserving. Figure 1 shows some orbits (i.e. for different initial points (q, p) the points $(q_n, p_n) = P^n(q, p)$ are plotted for $n \leq 1000$) of the standard map for different parameters κ. For $\kappa = 0$ an initial point (q, p) stays on the horizontal line and in q it rotates with frequency p. So for irrational p the corresponding line is filled densely. For $\kappa > 0$, the lines with rational p break up into an island-chain structure composed of (initially) stable orbits and their corresponding unstable (hyperbolic) partner. For small enough perturbation there are invariant (Kolmogorov–Arnold–Moser or short KAM) curves which are absolute barriers to the motion (for a more detailed discussion of these aspects the review [1]

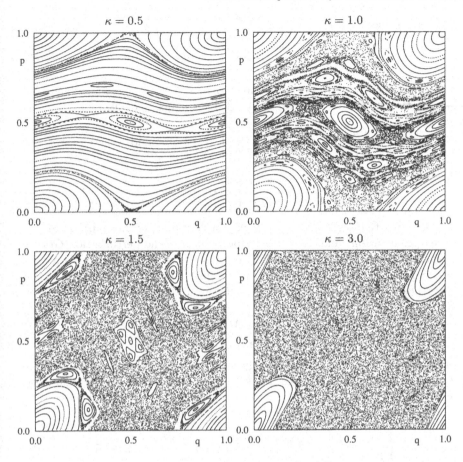

Fig. 1. Examples of orbits in the standard map for different parameters κ.

is a good starting point). For stronger perturbations, e.g. $\kappa = 1$ or $\kappa = 1.5$, the stochastic bands become larger and for even stronger perturbation (see the picture for $\kappa = 3.0$) there appears to be just one quite big stochastic region together with the elliptic island. The elliptic islands coexist with regions of irregular motion, therefore the standard map is an example of a so-called system with mixed phase space or, more briefly, a *mixed system*. Whether the motion in those stochastic regions is ergodic is one of the big unsolved problems, see [2] for a review on the coexistence problem. For some recent results on the classical dynamics of the standard map, in particular at large parameters, see [3–5].

An alternative way to specify a map $P : \mathbb{T}^2 \to \mathbb{T}^2$ is to use a generating function $S(q', q)$, from which the map is obtained by

$$p = -\frac{\partial S(q', q)}{\partial q} \qquad p' = \frac{\partial S(q', q)}{\partial q'} \; . \tag{4}$$

One easily checks that

$$S(q',q) = \frac{1}{2}(q-q')^2 + \frac{\kappa}{4\pi^2}\cos(2\pi q) \ , \tag{5}$$

is a generating function for the standard map (3).

Another important class are perturbed cat maps [6,7], like

$$\begin{pmatrix} q' \\ p' \end{pmatrix} = A \begin{pmatrix} q \\ p \end{pmatrix} + \kappa G(q) \begin{pmatrix} A_{12} \\ A_{22} \end{pmatrix} \qquad \mathrm{mod}\ 1 \ , \tag{6}$$

where

$$A = \begin{pmatrix} A_{11} & A_{12} \\ A_{21} & A_{22} \end{pmatrix} \tag{7}$$

is a matrix with integer entries (ensuring the continuity of the map), $\det A = 1$ (area preservation) and $\mathrm{Tr}\,A > 2$ (hyperbolicity). The perturbation $G(q)$ is a smooth periodic function on $[0,1[$. For $\kappa = 0$ the mapping is Anosov (see e.g. [8]), in particular it is ergodic and mixing. Moreover, following from the the Anosov theorem the map (6) is structurally stable, i.e. it stays Anosov as long

$$\kappa \le \kappa_{\mathrm{max}} := \frac{\sqrt{(\mathrm{Tr}\,A)^2 - 4} - \mathrm{Tr}\,A + 2}{2\max_q |G'(q)|\sqrt{1 + A_{22}^2}} \ ; \tag{8}$$

in particular the orbits are topologically conjugate to those of the unperturbed cat map. For larger parameters there are typically elliptic islands, so it becomes a mixed system.

A common choice for A and the perturbation is

$$\begin{pmatrix} q' \\ p' \end{pmatrix} = \begin{pmatrix} 2 & 1 \\ 3 & 2 \end{pmatrix} \begin{pmatrix} q \\ p \end{pmatrix} + \frac{\kappa}{2\pi}\cos(2\pi q) \begin{pmatrix} 1 \\ 2 \end{pmatrix} \qquad \mathrm{mod}\ 1 \ . \tag{9}$$

For $\kappa \le \kappa_{\mathrm{max}} = (\sqrt{3} - 1)/\sqrt{5} = 0.33\ldots$ the map is Anosov. The corresponding generating function is given by

$$S(q',q) = q'^2 - qq' + q^2 + \frac{\kappa}{4\pi^2}\sin(2\pi q) \ . \tag{10}$$

In Fig. 2a) one orbit for 20 000 iterations for the perturbed cat map (9) with $\kappa = 0.3$ is shown. The orbit appears to fill the torus in a uniform way, as it has to be asymptotically for almost all initial conditions because of the ergodicity of the map. For $\kappa = 6.5$ Fig. 2b) shows one orbit (20 000 iterates) in the irregular component and some orbits (1000 iterations) in the elliptic islands.

2.2 Quantization of Area-Preserving Maps

For the quantization of area–preserving maps exist several approaches, see for example [6,9–16]; a detailed account can be found in [17], and [18] provides

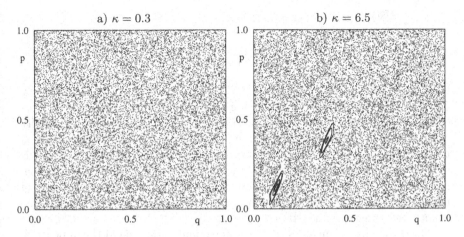

Fig. 2. Examples of orbits in the perturbed cat map (9) for $\kappa = 0.3$ and $\kappa = 6.5$.

a pedagogical introduction to the subject. First one has to find a suitable Hilbert space which incorporates the topology of the torus $\mathbb{T}^2$, i.e. the eigenfunctions in position and momentum have to fulfil

$$\psi(q+j) = e^{\frac{i}{\hbar}j\theta_2}\psi(q) \qquad ; \; j \in \mathbb{N} \qquad (11)$$

$$\widehat{\psi}(p+k) = e^{-\frac{i}{\hbar}k\theta_1}\widehat{\psi}(p) \qquad ; \; k \in \mathbb{N} \; . \qquad (12)$$

These conditions imply that Planck's constant $\hbar$ can only take the values $\hbar = \frac{1}{2\pi N}$ with $N \in \mathbb{N}$. Thus the semiclassical limit $\hbar \to 0$ corresponds to $N \to \infty$. The phases $(\theta_1, \theta_2) \in [0, 1[^2$ are at first arbitrary; for $\theta_1 = \theta_2 = 0$ one obtains periodic boundary conditions. For each N one has a Hilbert space $\mathcal{H}_N$ of finite dimension N. Observables $f \in C^\infty(\mathbb{T}^2)$ can be quantized analogous to the Weyl quantization to give an operator $\mathrm{Op}(f)$ on $\mathcal{H}_N$. Finally, a quantum map is a sequence of unitary operators U_N, $N \in \mathbb{N}$ on a Hilbert space $\mathcal{H}_N$. The quantum map is a quantization of a classical map P on $\mathbb{T}^2$, if the so–called Egorov property is fulfilled, i.e.

$$\lim_{N \to \infty} \|U_N^{-1}\mathrm{Op}(f)U_N - \mathrm{Op}(f \circ P)\| = 0 \qquad \forall f \in C^\infty(\mathbb{T}^2) \; . \qquad (13)$$

This means that semiclassically quantum time evolution and classical time evolution commute.

So the aim is to find for a given classical map a corresponding sequence of unitary operators. Unfortunately, this is not as straight forward as the quantization of Hamiltonian systems and a lot of information on this can be found in the above cited literature and references therein. One of the simplest approaches to determine U_N corresponding to a given area–preserving map uses its generating function to define

$$(U_N)_{j',j} := \langle q_{j'}|U_N|q_j\rangle$$

$$= \frac{1}{\sqrt{N}} \left|\frac{\partial^2 S(\tilde{q}',\tilde{q})}{\partial\tilde{q}'\partial\tilde{q}}\right|^{1/2}_{\tilde{q}'=q_{j'},\tilde{q}=q_j} \exp\left(2\pi\mathrm{i}NS\left(q_{j'},q_j\right)\right) \tag{14}$$

with $q_j = j/N$, $q_{j'} = j'/N$, where $j, j' = 0, 1, \ldots, N-1$. In the same way one may (and for certain maps which cannot represented in terms of $S(q',q)$ one has to) use other generating functions such as $S(p',p)$ or $S(q,p)$; usually these will lead to different eigenvalues and eigenfunctions. The question is to determine conditions on the generating function $S(q',q)$ such that U_N is unitary and fulfils the Egorov property (13). To my knowledge this question has not yet been fully explored, even though the quantum maps studied in the literature provide both examples and counterexamples. We will leave this as an interesting open question.

For the examples introduced before the quantization via (14) can be used. For the standard map we get

$$(U_N)_{j',j} = \frac{1}{\sqrt{N}} \exp\left[\frac{\mathrm{i}\pi}{N}(j'-j)^2 + \frac{\mathrm{i}\kappa N}{2\pi}\cos\left(\frac{2\pi}{N}j\right)\right] \tag{15}$$

with $j, j' = 0, \ldots, N-1$. A quantization of the standard map which takes the symmetries into account can be found in [19]. For the perturbed cat map (9) one gets using its generating function (10)

$$(U_N)_{j',j} = \frac{1}{\sqrt{N}} \exp\left(\frac{2\pi\mathrm{i}}{N}(j'^2 - j'j + j^2) + \mathrm{i}N\frac{\kappa}{2\pi}\sin(2\pi j/N)\right) . \tag{16}$$

For the unitary operator one has to solve the eigenvalue problem

$$U_N\psi_n = \lambda_n\psi_n \qquad \text{with } n = 0, \ldots, N-1, \ \psi_n \in \mathbb{C}^N . \tag{17}$$

Here λ_n is the n–th eigenvalue and the corresponding eigenvector ψ_n consists of N complex components, where N is the size of the unitary matrix U_N. Because of the unitarity of U_N the eigenvalues lie on the unit circle, i.e. $|\lambda_n| = 1$.

Let us discuss some of the numerical aspects relevant for finding the solutions of (17) without going into implementation specific details (see the appendix and [20] for an implementation using Python).

Computing the eigenvalues of (17) consists of two main steps

- Setting up the matrix U_N:
  The computational effort increases proportional to N^2 (unless each matrix element requires further loops) as we have to fill the N^2 matrix elements. The memory requirement to store U_N is $16\,N^2$ Bytes (for a IEEE-compliant machine a double precision floating point number requires 8 Bytes; as we have both real and imaginary part we end up with 16 Bytes per matrix element).

– Computing the eigenvalues:
The computational effort for the matrix diagonalization (typically) scales like N^3.
Usually one will use a black-box routine such as one from the NAG-library [21] or from LAPACK [22]. To my knowledge there are no routines which make use of the fact that the matrix U_N is unitary so we may for example use the NAG routine F02GBF or the LAPACK routine ZGEES (or the more recent routine ZGEEV which is faster for larger matrices, e.g. $N \geq 500$) which compute all eigenvalues of a complex matrix.
For certain maps specific optimizations are possible, see e.g. [19] for the standard map. For this type of mapping a different approach employing a combination of fast Fourier transform and Lanczos method reduces the computational effort to $N^2 \ln N$ [23].

After successful compilation and running of the program it is useful to see whether the eigenvalues really lie on the unit circle. In Fig. 3 this is illustrated for $N = 200$ and the standard map with $\kappa = 1.5$. For small N the running times of the program for setting up the matrix U_N and its diagonalization is just a matter of minutes. For example on an Intel Pentium III processor with 666 MHz one needs just 6 minutes to compute the eigenvalues of (17) when $N = 1000$. However, for $N = 3000$ already 140 MB of RAM are required to store U_N and the computing time increases to 6 hours. Depending on

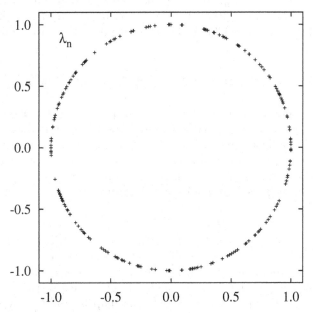

Fig. 3. Plotting the eigenvalues of U_N allows to check the numerical implementation and the unitarity of U_N; the picture shows for $N = 200$ and $\kappa = 1.5$ the eigenvalues λ_n for the quantized standard map (15).

available memory, computing power, patience and motivation one may use larger values of N.

Let us conclude this part with a more technical remark: In addition to the choice of computer language, compiler, optimizations and algorithm there is one very important component for achieving good performance when doing numerical linear algebra computations: the BLAS (Basic Linear Algebra Subprograms). Libraries such as LAPACK defer all the basics tasks like adding vectors, vector–matrix or matrix–matrix multiplication to the BLAS such that highly optimized (machine-specific) BLAS routines should be used. Most hardware vendors provide these (of differing quality). Recently the software system ATLAS (Automatically Tuned Linear Algebra Software) [24] has been introduced which generates a machine dependent optimized BLAS library. For some computers ATLAS-based BLAS can be even faster than the vendor supplied ones!

2.3 Eigenvalue Statistics

One central research line in quantum chaos is the investigation of spectral statistics. It has been conjectured [25] that for generic chaotic systems the eigenvalue statistics can be described by random matrix theory, whereas generic integrable systems should follow Poissonian statistics [26]. To study the eigenvalue statistics for quantum maps one considers the eigenphases $\varphi_n \in [0, 2\pi[$, defined by $\lambda_n = \mathrm{e}^{\mathrm{i}\varphi_n}$ (in the following we will also call φ_n levels in analogy to the energy levels for the Schrödinger equation). The simplest statistics is the nearest neighbour level spacing distribution $P(s)$ which is the distribution of the spacings

$$s_n := \frac{N}{2\pi}(\varphi_{n+1} - \varphi_n) \qquad \text{with } n = 0, \ldots, N-1 \quad \text{and } \varphi_N := \varphi_0 \ .$$

The factor $\frac{N}{2\pi}$ ensures that the average of all spacings s_n is 1. To compute the distribution practically one chooses a division of the interval $[0, 10]$ (usually this interval is sufficient, but more precisely the upper limit is determined by the largest s_n) into b bins and determines the fraction of spacings s_n falling into the corresponding bins. If N is too small it is better to consider instead of $P(s)$ the corresponding cumulative distribution

$$I(s) := \frac{\#\{n \mid s_n \leq s\}}{N} \tag{18}$$

which avoids the binning and results in a smoother curve.

Fig. 4 shows for the perturbed cat map (9) with $\kappa = 0.3$ the level spacing distribution $P(s)$ and the cumulative level spacing distribution $I(s)$ for $N = 3001$. For this parameter value κ the map is still Anosov so one expects that the correlations of the eigenphases follow random matrix theory; in particular because the perturbation should break up the number theoretical degeneracies which lead to non-generic spectral statistics for the cat maps

(a)

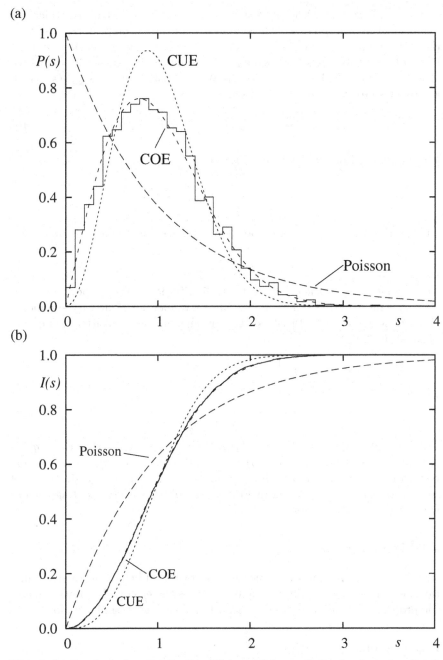

Fig. 4. (a) Level spacing distribution $P(s)$ and (b) cumulative level spacing distribution $I(s)$ for the perturbed cat map (9) with $\kappa = 0.3$ and $N = 3001$.

at $\kappa = 0$ [27, 28]. In [29, 30] it is shown that for all perturbations which are just a shear in position one of the symmetries of the cat map survives, so that the statistics are expected to be described by the circular orthogonal ensemble (COE). In the limit $N \to \infty$ this is the same as the Gaussian orthogonal ensemble (GOE). In Fig. 4 we show the Wigner distribution $P_{\text{Wigner}}(s)$ which is very close to the COE distribution,

$$P_{\text{COE}}(s) \approx P_{\text{Wigner}}(s) = \frac{\pi}{2} s \exp\left(-\frac{\pi}{4} s^2\right) . \tag{19}$$

and for comparison the CUE distribution

$$P_{\text{CUE}}(s) \approx \frac{32}{\pi^2} s^2 \exp\left(-\frac{4}{\pi} s^2\right) \tag{20}$$

and the Poisson distribution (expected for generic integrable systems)

$$P_{\text{Poisson}}(s) = e^{-s} . \tag{21}$$

The agreement with the expected COE distribution is very good.

A specific example, which breaks the above mentioned unitary symmetry and thus leads to CUE statistics, uses two shears, one in position and one in momentum [29],

$$\begin{pmatrix} q' \\ p' \end{pmatrix} = (A \circ P_q \circ P_p) \begin{pmatrix} q \\ p \end{pmatrix} , \tag{22}$$

where

$$A = \begin{pmatrix} 12 & 7 \\ 41 & 24 \end{pmatrix} \tag{23}$$

and $P_q(q, p) = (q + \kappa_q G(p), p)$, $P_p(q, p) = (q, p + \kappa_p F(q))$ with $F(q) = \frac{1}{2\pi}(\sin(2\pi q) - \sin(4\pi q))$ and $G(p) = \frac{1}{2\pi}(\sin(4\pi q) - \sin(2\pi q))$. For the corresponding quantum map with $\kappa_p = \kappa_q = 0.012$ and $N = 3001$ the level spacing distribution is shown in Fig. 5. One observes very good agreement with the CUE distribution.

2.4 Eigenfunctions

Another interesting question concerns the statistical behaviour of eigenfunctions, and more specifically for quantum maps the eigenvector statistics and the properties of phase space representations like the Husimi function.

Eigenvector Distributions

Consider an eigenvector ψ of a quantum map given by the N numbers $c_j \in \mathbb{C}$, $j = 0, ..., N - 1$. The distribution $P(\psi)$ is given (similarly to the level spacing distribution) by

(a)

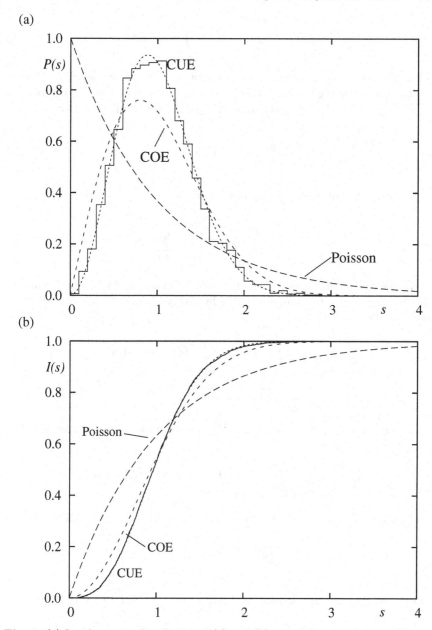

(b)

Fig. 5. (a) Level spacing distribution $P(s)$ and (b) cumulative level spacing distribution $I(s)$ for the perturbed cat map (22) with $\kappa_p = \kappa_q = 0.012$ and $N = 3001$.

$$\frac{1}{N}\#\{a \leq |c_j|^2 \leq b\} = \int_a^b P(\psi)\mathrm{d}\psi \ . \tag{24}$$

Let us first discuss the corresponding random matrix results (see e.g. [31,32]). For the COE the eigenvectors can be chosen to be real and the coefficients c_j, $j = 0, \ldots, N-1$, only have to obey the normalization condition

$$\sum_{j=0}^{N-1} c_j^2 = 1 \qquad \text{with } c_j \in \mathbb{R} \ . \tag{25}$$

Thus the joint probability for an eigenvector $\boldsymbol{c} = (c_0, \ldots, c_{N-1}) \in \mathbb{R}^N$ is

$$P_N^{\mathrm{COE}}(\boldsymbol{c}) = \frac{\Gamma(N/2)}{\pi^{N/2}}\delta\left(1 - \sum_{j=0}^{N-1} c_j^2\right) \ , \tag{26}$$

where the prefactor ensures normalization. So the probability of one component to have a specific value y is given by integrating $P_N^{\mathrm{COE}}(\boldsymbol{c})$ over all other components,

$$\begin{aligned} P_N^{\mathrm{COE}}(y) &= \int \delta(y - c_0^2) P_N^{\mathrm{COE}}(\boldsymbol{c}) \ \mathrm{d}c_0 \cdots \mathrm{d}c_{N-1} \\ &= \frac{1}{\sqrt{\pi y}} \frac{\Gamma(N/2)}{\Gamma((N-1)/2)}(1-y)^{(N-3)/2} \ . \end{aligned} \tag{27}$$

The mean of $P_N^{\mathrm{COE}}(y)$ is $\int_0^1 y P_N^{\mathrm{COE}}(y) \ \mathrm{d}y = 1/N$. So using the rescaling $\eta = yN$ gives

$$P_N^{\mathrm{COE}}(\eta) = \frac{1}{\sqrt{\pi N \eta}} \frac{\Gamma(N/2)}{\Gamma((N-1)/2)}(1 - \eta/N)^{(N-3)/2} \ . \tag{28}$$

In the limit of large N one gets the so-called Porter-Thomas distribution [33]

$$P_N^{\mathrm{COE}}(\eta) = \frac{1}{\sqrt{2\pi\eta}} \exp(-\eta/2) \ , \tag{29}$$

and the corresponding cumulative distribution $I(y) = \int_0^y P(y') \ \mathrm{d}y'$ reads

$$I(\eta) = \mathrm{erf}\left(\sqrt{\eta/2}\right) \ . \tag{30}$$

Figure 6 shows an example for the eigenvector distribution of an eigenstate of the perturbed cat map (9) with $\kappa = 0.3$ and $N = 1597$. There is good agreement with the expected COE distribution, (29), shown as dashed line.

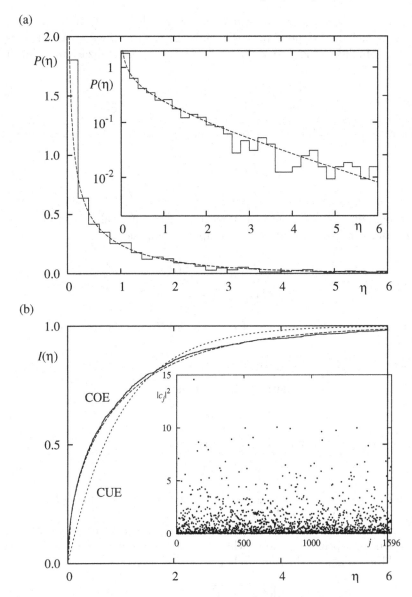

Fig. 6. (a) Eigenvector distribution for the perturbed cat map (9) with $N = 1597$ and $\kappa = 0.3$. In comparison with the asymptotic COE distribution, (28), dashed line. The inset shows the same curves in a log-normal plot. In (b) the cumulative distribution is shown and in the inset a plot of the absolute value of the components $c_j^{(n)}$, $j = 0, \ldots, N-1$ of the corresponding eigenvector $\psi_{n=20}$ is displayed.

Finally, let us consider again the map (22) which shows CUE level statistics. From this one would expect that also the eigenvector statistics follows the CUE. Similar to the case of the COE one has the normalization condition

$$\sum_{j=0}^{N-1} |c_j|^2 = 1 \qquad \text{with } c_j \in \mathbb{C} \ . \tag{31}$$

Thus the joint probability for an eigenvector $\boldsymbol{c} = (c_0, \ldots, c_{N-1}) \in \mathbb{C}^N$ reads

$$P_N^{\text{CUE}}(\boldsymbol{c}) = \frac{\Gamma(N)}{\pi^N} \delta \left(1 - \sum_{j=0}^{N-1} |c_j|^2 \right) \ . \tag{32}$$

The probability of one component to have a specific value y is given by integrating $P_N^{\text{CUE}}(\boldsymbol{c})$ over all other (complex) components,

$$
\begin{aligned}
P_N^{\text{CUE}}(y) &= \int \delta(y - |c_0|^2) P_N^{\text{COE}}(\boldsymbol{c}) \, \mathrm{d}^2 c_0 \cdots \mathrm{d}^2 c_{N-1} \\
&= (N-1)(1-y)^{N-2} \ .
\end{aligned}
\tag{33}
$$

Again as for the COE, the mean of $P_N^{\text{CUE}}(y)$ is $1/N$ and the rescaling $\eta := yN$ leads to

$$P_N^{\text{CUE}}(\eta) = \frac{N-1}{N} \left(1 - \frac{\eta}{N} \right)^{N-2} \tag{34}$$

which has mean 1. In the large N limit we have

$$P^{\text{CUE}}(\eta) = \exp(-\eta) \ . \tag{35}$$

and the cumulative distribution simply is

$$I^{\text{CUE}}(\eta) = 1 - \exp(-\eta) \ . \tag{36}$$

Figure 7 shows $P(\eta)$ for one eigenvector of the perturbed cat map (22). There is good agreement with $P^{\text{CUE}}(\eta)$.

A different distribution is obtained for unperturbed cat maps: for certain subsequences of prime numbers (which depend on the map) the distribution of $\eta = \frac{1}{2} \text{Re}\, \psi$ tends to the semicircle law,

$$P(\psi) = \begin{cases} \frac{2}{\pi} \sqrt{1 - \eta^2} & \text{for } \eta \leq 1 \\ 0 & \text{for } \eta > 1 \ , \end{cases} \tag{37}$$

see [34] for details (see also [35]). In Fig. 8 we show an example of an eigenstate with $N = 1597$ for the quantum map corresponding to the map (9) with $\kappa = 0$. For this N the map fulfils the conditions of [34] and one observes a nice semicircle distribution of the eigenvector. However, it seems that the approach to the asymptotic distribution is slower than for the case of the random matrix situations.

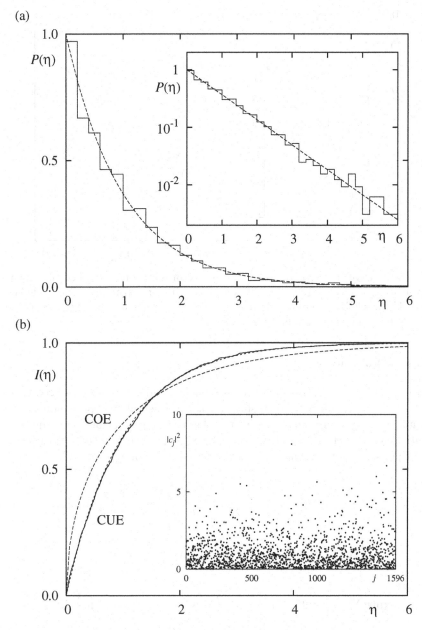

Fig. 7. (a) Eigenvector distribution of an eigenvector for the perturbed cat map (22) with $N = 1597$ and $\kappa_p = \kappa_q = 0.012$ is shown in comparison with the asymptotic CUE distribution, (35), dashed line. The inset shows the same curves in a log-normal plot. In (b) the corresponding cumulative distributions are shown and in the inset a plot of the absolute value of the components $c_j^{(n)}$, $j = 0, \ldots, N - 1$ of the corresponding eigenvector $\psi_{n=2}$ is displayed.

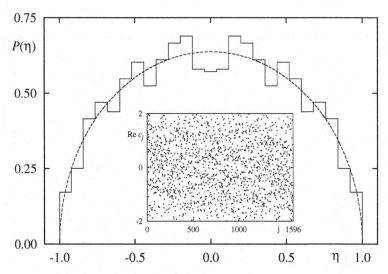

Fig. 8. Eigenvector distribution of an eigenvector for the unperturbed (i.e. $\kappa = 0$) cat map (9) with $N = 1597$. This is compared with the asymptotic semicircle law, (37). The inset shows the corresponding eigenvector (compare with the eigenvectors shown in the previous two figures).

Husimi Functions

A different representation of eigenstates is to consider a phase space representation, like for example the Husimi function, which allows for a more direct comparison with the structures for the classical map. Without going into the mathematical details, the Husimi representation is obtained by projecting the eigenstate onto a coherent state centered in a point $(q, p) \in \mathbb{T}^2$,

$$H_n(q, p) = |\langle C_{q,p} | \psi_n \rangle|^2 = \left| \sum_{j=0}^{N-1} \langle C_{q,p} | q_j \rangle \langle q_j | \psi \rangle \right|^2 = \left| \sum_{j=0}^{N-1} \langle C_{q,p} | q_j \rangle c_j \right|^2$$

$$= \left| \sum_{j=0}^{N-1} (2N)^{1/4} \exp\left(-\pi N(q^2 - ipq)\right) \right.$$

$$\left. \exp(\pi N(-q_j^2 + 2(q - ip)q_j)) \vartheta_3\left(i\pi N \left(q_j - \frac{i\theta_1}{N} - q + ip \right) \Big| iN \right) c_j \right|^2 .$$

Here $q_j = \frac{1}{N}(\theta_2 + j)$, $j = 0, \ldots, N - 1$ and $\vartheta_3(Z|\tau)$ is the Jacobi-Theta function,

$$\vartheta_3(Z|\tau) = \sum_{n \in \mathbb{Z}} e^{i\pi\tau n^2 + 2inZ} , \qquad \text{with } Z, \tau \in \mathbb{C}, \ \text{Im}(\tau) > 0 . \tag{39}$$

The coefficients c_j are the components of the eigenvector ψ_n in the position representation as obtained from the diagonalization of U_N (for other generating functions than the one used in (14) one has to adapt (38)).

If one wants to compute a Husimi function on a grid of $N \times N$ points on $\mathbb{T}^2$ the computational effort grows with N^3. So for computing all Husimi function of a quantum map for a given N the computational effort grows with N^4. Already for moderate N this can be quite time-consuming, but even more importantly, usually one also wants to store all these Husimi functions on the hard-disk which limits the accessible range of N. Sometimes a smaller grid, e.g. of size $10\sqrt{N} \times 10\sqrt{N}$ can be sufficient which reduces the growth of the computational effort to N^2 for a single Husimi function and to N^3 for all Husimi functions at a given N. Even then one still needs $800\,N^2$ Bytes to store these on the hard-disk. For example for $N = 1600$ this roughly leads to 2 GB of data and for $N = 3000$ one needs approximately 7 GB. However, there are also cases where a finer grid, e.g. $2N \times 2N$ is necessary.

Theoretically one expects that for $N \to \infty$ the Husimi functions concentrate on those regions in phase space which are invariant under the map (this follows from the Egorov property). So for ergodic systems the expectation is that (in the weak sense)

$$H_n(q,p) \to 1 \quad \text{with } n = 0, ..., N-1 \text{ as the matrix size } N \to \infty \ . \quad (40)$$

The precise formulation of this statement is the contents of the quantum ergodicity theorem for maps [36] (see [37] for the case of discontinuous maps). The quantum ergodicity theorem only makes a statement about a subsequence of density one (i.e. almost all states) which for example leaves space for scars, i.e. eigenstates localized on unstable periodic orbits. For systems with mixed phase space one (asymptotically) expects localization in the stochastic region(s) and on the tori in the elliptic regions.

In Fig. 9a) we show for the perturbed cat map with $\kappa = 0.3$ the Husimi function for the same eigenstate as in Fig. 6. As expected it shows a quite uniform distribution (of course with the usual fluctuations). In contrast for $\kappa = 6.5$ there are eigenstates such as the one shown in Fig. 9b) which localizes on the elliptic island (compare with Fig. 2).

In some sense more interesting are the Husimi functions for mixed systems as the classical dynamics shows more structure. In Fig. 10 we show some examples for the standard map with $\kappa = 3.0$. Figure 10a) shows a Husimi function which is spread out in the irregular component. In contrast in b) the Husimi function localizes on a torus around the elliptic fixpoint. The Husimi function in c) shows quite strong localization around the small elliptic island of a periodic orbit with period 4. This island is so small that it is not visible in Fig. 1. Therefore, the Husimi function displayed in Fig. 10d) indicates that the region of 'influence' of this island is much larger than the area of the island. This region is also visible in the Husimi function in Fig. 10a), as the irregular state has a very small probability in the regions around these

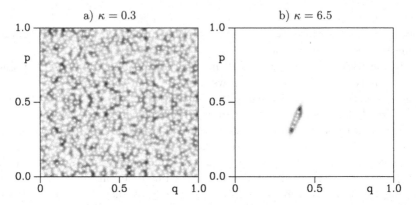

Fig. 9. In a) a Husimi function $H_n(q,p)$ of the perturbed cat map (9) with $\kappa = 0.3$ is plotted which shows the expected 'uniform' distribution. Here black corresponds to large values of $H_n(q,p)$. In b) for $\kappa = 6.5$ a state localizing on one of the elliptic islands is shown (compare with Fig. 2).

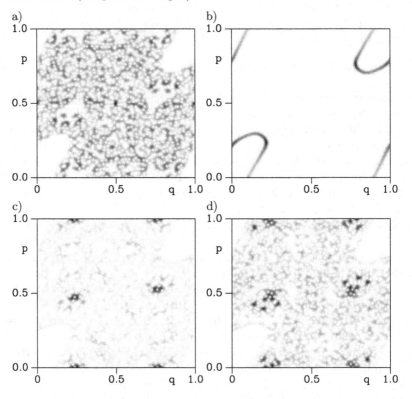

Fig. 10. Examples of Husimi functions for the standard map with $\kappa = 3.0$ and $N = 1600$.

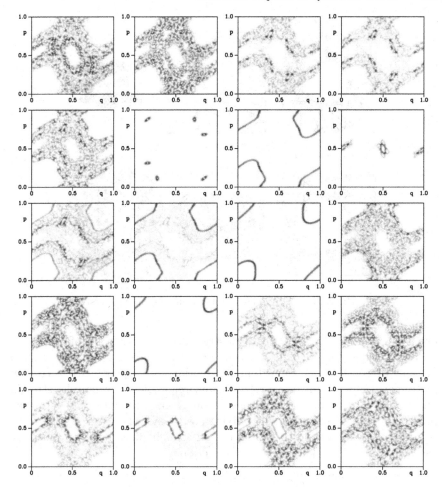

Fig. 11. Examples of Husimi functions $H_n(q, p)$ for the standard map with $\kappa = 1.5$ and $N = 1600$ for $n = 0, \ldots, 19$. (Compare with Fig. 1.)

islands. A longer sequence of Husimi functions for the standard map with $\kappa = 1.5$ shown in Fig. 11 illustrates the different types of localized states (compare with Fig. 1).

3 Billiards

3.1 Classical Billiards

A two-dimensional Euclidean billiard is given by the free motion of a point particle in some domain $\Omega \subset \mathbb{R}^2$ with elastic reflections at the boundary $\partial\Omega$. Depending on the boundary one obtains completely different dynamical

Integrable systems

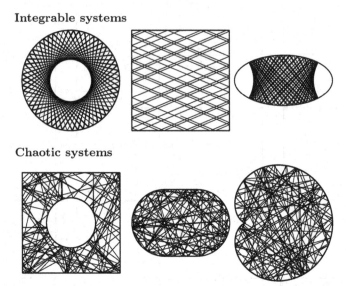

Chaotic systems

Fig. 12. Billiard dynamics in integrable and chaotic systems.

behaviour, see Fig. 12 where this is illustrated by showing orbits of billiards in a circle, a square and an ellipse, which are all integrable giving rise to regular motion. In contrast the Sinai billiard (motion in a square with a circular scatterer), the stadium billiard (two semicircles joined by parallel straight lines) and the cardioid billiard show strongly chaotic motion (they are all proven to be hyperbolic, ergodic, mixing and K-systems).

As the motion inside the billiard is on straight lines it is convenient to use the boundary to define a Poincaré section,

$$\mathcal{P} := \{(s,p) \mid s \in [0, |\partial\Omega|], \ p \in [-1,1]\} \ . \tag{1}$$

Here s is the arclength along $\partial\Omega$ and $p = \langle v, T(s)\rangle$ is the projection of the unit velocity vector v after the reflection on the unit tangent vector $T(s)$ in the point $s \in \partial\Omega$. The Poincaré map is then given by

$$P : \mathcal{P} \to \mathcal{P}$$
$$\xi = (s,p) \mapsto \xi' = (s',p') \ ,$$

i.e. for a given point $\xi = (s,p)$ one considers the ray starting at the point $r(s) \in \partial\Omega$ in the direction specified by p and determines the first intersection with the boundary, leading to the new point $\xi' = (s',p')$. Explicitly, the Cartesian components of the unit velocity v of a point particle starting on $\partial\Omega$ at $r(s)$ are determined by the angle $\beta \in [-\pi/2, \pi/2]$ measured with respect to the inward pointing normal $N = (-T_y, T_x)$. The velocity in the T, N coordinate system is denoted by $(p,n) = (\sin\beta, \cos\beta)$, so that in Cartesian coordinates

$$\boldsymbol{v} = (v_x, v_y) = \begin{pmatrix} T_x & N_x \\ T_y & N_y \end{pmatrix} (p, n)$$
$$= \left(T_x p + N_x \sqrt{1 - p^2}, T_y p + N_y \sqrt{1 - p^2} \right) \ . \tag{2}$$

Starting in the point $\boldsymbol{r}(s) \in \partial\Omega$ in the direction $\boldsymbol{v}$, the ray $\boldsymbol{r} + t\boldsymbol{v}$ intersects $\partial\Omega$ at some point $\boldsymbol{r}' = (x', y')$. If the boundary is determined by the implicit equation

$$F(x, y) = 0 \ , \tag{3}$$

the new point $\boldsymbol{r}'$ can be determined by solving

$$F(x + tv_x, y + tv_y) = 0 \ . \tag{4}$$

For non-convex billiards there are points $\xi = (s, p) \in \mathcal{P}$ for which there is more than one solution (apart from $t = 0$); obviously the one with the smallest $t > 0$ has to be chosen. The condition (3) can be used to remove the $t = 0$ solution analytically from (4). If F is a polynomial in x and y this allows to reduce the order of (4) by one. This approach has for example been used for the cardioid billiard leading to a cubic equation for t, see [38] for details. From the solution t one gets the coordinates $(x', y') = (x, y) + t\boldsymbol{v}$ which have to be converted (in a system dependent way) to the arclength coordinate s' (in many practical applications there is a more suitable internal variable, for example the polar angle etc.). The corresponding new projection of the momentum is given by $p' = -\langle \boldsymbol{v}, \boldsymbol{T}(s') \rangle$.

3.2 Quantum Billiards

For a classical billiard system the associated quantum billiard is given by the stationary Schrödinger equation (in units $\hbar = 2m = 1$)

$$-\Delta\psi_n(\boldsymbol{q}) = E_n\psi_n(\boldsymbol{q}) \ , \quad \boldsymbol{q} \in \Omega \tag{5}$$

with (for example) Dirichlet boundary conditions, i.e. $\psi_n(\boldsymbol{q}) = 0$ for $\boldsymbol{q} \in \partial\Omega$. Here Δ denotes the Laplace operator, which reads in two dimensions

$$\Delta = \left(\frac{\partial^2}{\partial q_1^2} + \frac{\partial^2}{\partial q_2^2} \right) \ . \tag{6}$$

In the Schrödinger representation the state of a particle is described in configuration space by a wave function $\psi \in L^2(\Omega)$, where $L^2(\Omega)$ is the Hilbert space of square integrable functions on Ω. The interpretation of ψ is that $\int_D |\psi(\boldsymbol{q})|^2 \, \mathrm{d}^2 q$ is the probability of finding the particle inside the domain $D \subset \Omega$.

Due to the compactness of Ω, the quantal energy spectrum $\{E_n\}$ is purely discrete and can be ordered as $0 < E_1 \leq E_2 \leq E_3 \leq \dots$. The eigenfunctions can be chosen to be real and to form an orthonormal basis of $L^2(\Omega)$,

$$\langle \psi_n | \psi_m \rangle := \int_{\Omega} \psi_n(\boldsymbol{q}) \psi_m(\boldsymbol{q}) \, \mathrm{d}^2 q = \delta_{mn} \ .$$

The mathematical problem defined by (5) is the well-known eigenvalue problem of the Helmholtz equation, which for example also describes a vibrating membrane or flat microwave cavities. For some simple domains Ω it is possible to solve (5) analytically. For example for the billiard in a rectangle with sides a and b the (non-normalized) eigenfunctions are given by $\psi_{n_1,n_2}(\boldsymbol{q}) = \sin(\pi n_1 q_1/a) \sin(\pi n_2 q_2/b)$ with corresponding eigenvalues $E_{n_1,n_2} = \pi^2(n_1^2/a^2 + n_2^2/b^2)$ and $(n_1, n_2) \in \mathbb{N}^2$. For the billiard in a circle the eigenfunctions are given in polar-coordinates by $\psi_{mn}(r, \varphi) = J_m(j_{mn}r) \exp(\mathrm{i}m\varphi)$, where j_{mn} is the n-th zero of the Bessel function $J_m(x)$ and $m \in \mathbb{Z}$, $n \in \mathbb{N}$. However, in general no analytical solutions of (5) exist so that numerical methods have to be used to compute eigenvalues and eigenfunctions.

The spectral staircase function $N(E)$ (integrated level density)

$$N(E) := \#\{n \mid E_n \leq E\} \tag{7}$$

counts the number of energy levels E_n below a given energy E. $N(E)$ can be separated into a mean smooth part $\overline{N}(E)$ and a fluctuating part

$$N(E) = \overline{N}(E) + N_{\mathrm{fluc}}(E) \ . \tag{8}$$

For two-dimensional billiards, $\overline{N}(E)$ is given by the generalized Weyl formula [39]

$$\overline{N}(E) = \frac{\mathcal{A}}{4\pi} E - \frac{\mathcal{L}}{4\pi} \sqrt{E} + \mathcal{C} + \dots \ , \tag{9}$$

where $\mathcal{A}$ denotes the area of the billiard, and $\mathcal{L} := \mathcal{L}^- - \mathcal{L}^+$, where $\mathcal{L}^-$ and $\mathcal{L}^+$ are the lengths of the pieces of the boundary $\partial\Omega$ with Dirichlet and Neumann boundary conditions, respectively. The constant $\mathcal{C}$ takes curvature and corner corrections into account.

The simplest quantity is the δ_n-statistics, which is obtained from the fluctuating part of the spectral staircase evaluated at the unfolded energy eigenvalues $x_n := \overline{N}(E_n)$

$$\delta_n := N_0(E_n) - \overline{N}(E_n) = n - \frac{1}{2} - x_n \ , \tag{10}$$

where

$$N_0(E) := \lim_{\epsilon \to 0} \frac{N(E + \epsilon) + N(E - \epsilon)}{2} \ . \tag{11}$$

The quantity δ_n is a good measure for the completeness of a given energy spectrum. For a complete spectrum δ_n, or equivalently $N_{\mathrm{fluc}}(x)$, should fluctuate around zero. Figure 13a) shows $N_{\mathrm{fluc}}(x)$ for the stadium billiard, which indeed fluctuates around zero. In addition there is an overall modulation of

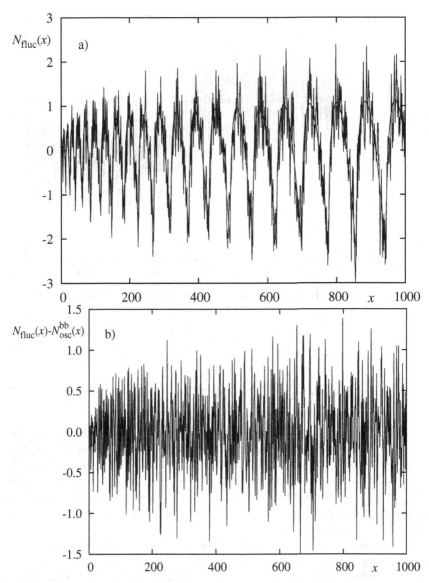

Fig. 13. Plot of $N_{\text{fluc}}(x)$ for the stadium billiard ($a = 1.8$, odd-odd symmetry) together with the contribution from the bouncing ball orbits, dashed line, see (12). In b) the fluctuating part after subtraction of the contribution of the bouncing ball orbits is shown.

$N_{\text{fluc}}(x)$ which is caused by the bouncing ball orbits. They form a one parameter family of periodic orbits having perpendicular reflections at the two parallel walls (of length a, see Fig. 15) of the stadium. The contribution of these orbits to the spectral staircase function reads [40]

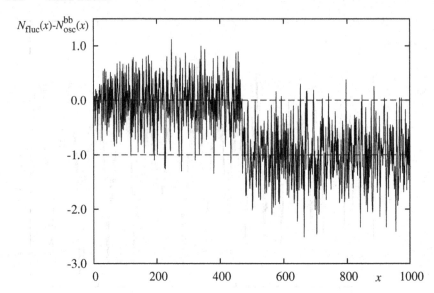

Fig. 14. Detection of missing levels using the δ_n-statistics.

$$N_{\text{fluc}}^{\text{bb}}(E) = \frac{a}{\pi} \sum_{n=1}^{\infty} \sqrt{E - E_n^{\text{bb}}} \; \Theta\left(\sqrt{E} - \sqrt{E_n^{\text{bb}}}\right) - \left(\frac{a}{4\pi}E - \frac{1}{2\pi}\sqrt{E}\right) \quad (12)$$

$$= \frac{a}{2\sqrt{\pi^3}} \, E^{\frac{1}{4}} \sum_{n=1}^{\infty} \frac{1}{n^{\frac{3}{2}}} \, \cos\left(2an\sqrt{E} - \frac{3\pi}{4}\right) \;, \quad (13)$$

where $E_n^{\text{bb}} = \pi^2 n^2$ are the eigenvalues of a particle in a one-dimensional box of length 1, and Θ is the Heaviside step function. Subtracting $N_{\text{osc}}^{\text{bb}}(x)$ from $N_{\text{fluc}}(x)$ removes the additional oscillation, see Fig. 13b). If an eigenvalue is missing this is clearly visible by a 'jump' of δ_n in comparison to points fluctuating around 0, see Fig. 14 for an example where one eigenvalue has been removed 'by-hand'. Clearly, the energy interval in which a level is missing can be estimated from the plot.

In the same way as for quantum maps one can study the level spacing distribution and more complicated statistics, like the number variance, n-point correlation functions etc., see for example [41, 42] for some further examples for the cardioid billiard.

3.3 Computing Eigenvalues and Eigenfunctions for Quantum Billiards

There exist several numerical methods to solve the Helmholtz equation inside a domain $\Omega \subset \mathbb{R}^2$,

$$\Delta\psi(\mathbf{q}) + k^2\psi(\mathbf{q}) = 0 \;, \qquad \mathbf{q} \in \Omega\backslash\partial\Omega \;, \quad (14)$$

with Dirichlet boundary conditions

$$\psi(\boldsymbol{q}) = 0 \ , \qquad \boldsymbol{q} \in \partial\Omega. \tag{15}$$

For a good review on the determination of the eigenvalues of (14) see [43], which however does not cover finite element methods or boundary integral methods. Additionally, in the context of quantum chaos the plane wave decomposition [44] (see also [45] for a detailed description of the method), the scattering approach, see e.g. [46–48], and more recently the scaling method [49], are commonly used.

   Here I will give a sketch of the derivation of the boundary integral method and discuss in more detail the numerical implementation. The boundary integral method reduces the problem of solving the two-dimensional Helmholtz equation (14) to a one-dimensional integral equation, see e.g. [50–65] and references therein. Of course, the general approach also applies to higher dimensions but we will only discuss the two-dimensional case. For studies of three-dimensional systems by various methods see e.g. [66–70]. Boundary integral methods are also used in many other areas so that it is impossible to give a full account. For example they are also commonly used in acoustics, see e.g. [71] and the detailed list of references therein. Finally, the boundary integral method provides a starting point to derive the Gutzwiller trace formula, see e.g. [64, 72–75].

Boundary Integral Equation

Let $G(\boldsymbol{q}, \boldsymbol{q}')$ be a Green function of the inhomogeneous equation, i.e.

$$(\Delta + k^2)G_k(\boldsymbol{q}, \boldsymbol{q}') = \delta(\boldsymbol{q} - \boldsymbol{q}') \ . \tag{16}$$

Considering the integral over Ω of the difference $\psi(\boldsymbol{q}')\cdot(16) - G_k(\boldsymbol{q}, \boldsymbol{q}')\cdot(14)$ one obtains

$$\int_\Omega [\psi(\boldsymbol{q}')\Delta' G_k(\boldsymbol{q}, \boldsymbol{q}') - G_k(\boldsymbol{q}, \boldsymbol{q}')\Delta'\psi(\boldsymbol{q}')] \ \mathrm{d}^2\boldsymbol{q}'$$
$$= \int_\Omega \psi(\boldsymbol{q}')\delta(\boldsymbol{q} - \boldsymbol{q}') \ \mathrm{d}^2\boldsymbol{q}' \ . \tag{17}$$

Using the second Green theorem gives the Helmholtz representation

$$\oint_{\partial\Omega} \left[\psi(\boldsymbol{q}')\frac{\partial G_k}{\partial n'}(\boldsymbol{q}, \boldsymbol{q}') - G_k(\boldsymbol{q}, \boldsymbol{q}')\frac{\partial\psi}{\partial n'}(\boldsymbol{q}') \right] \ \mathrm{d}s'$$
$$= \begin{cases} \psi(\boldsymbol{q}) \ ; & \boldsymbol{q} \in \Omega \setminus \partial\Omega \\ \frac{1}{2}\psi(\boldsymbol{q}) \ ; & \boldsymbol{q} \in \partial\Omega \\ 0 \ ; & \text{else} \end{cases} \tag{18}$$

Here $q' \equiv q(s')$ and $\frac{\partial}{\partial n'} = n(s')\nabla$ with $n(s) = (q_2'(s), -q_1'(s))$ denoting the outward pointing normal vector, where $(q_1(s), q_2(s))$ is a parametrization of the billiard boundary $\partial\Omega$ in terms of the arclength s, oriented counterclockwise. Special care has to be taken to obtain the result for $q \in \partial\Omega$, see e.g. [51, 74]. (When q is in a corner of the billiard the factor $\frac{1}{2}$ has to be replaced by $\frac{\theta}{2\pi}$, where θ is the (inner) angle of the corner.) For Dirichlet boundary conditions one obtains

$$\oint_{\partial\Omega} u(s')G_k(q, q')\,\mathrm{d}s' = 0 , \qquad q \in \partial\Omega , \tag{19}$$

where

$$u(s) := \frac{\partial}{\partial n}\psi(q(s)) := n(s)\nabla\psi(q(s)) := n(s) \lim_{\substack{q' \to q(s) \\ q' \in \Omega\backslash\partial\Omega}} \nabla\psi(q') \tag{20}$$

is the normal derivative function of ψ.

In two dimensions a Green function for a free particle is given by the Hankel function of first kind

$$\begin{aligned} G_k(q, q') &= -\frac{\mathrm{i}}{4} H_0^{(1)}(k|q - q'|) \\ &= -\frac{\mathrm{i}}{4} [J_0(k|q - q'|) + \mathrm{i}\,Y_0(k|q - q'|)] . \end{aligned} \tag{21}$$

Since $H_0^{(1)}(z) \sim \frac{\mathrm{i}}{\pi} \ln z$ for $z \to 0$, the Green function $G_k(q, q')$ diverges logarithmically such that it is more convenient to derive an integral equation whose kernel is free of this singularity. To that end one (formally) applies the normal derivative $\frac{\partial}{\partial n}$ to (18). More carefully one has to consider a jump relation for the normal derivative function, see e.g. [51, 74]. The result is

$$u(s) = -2 \oint_{\partial\Omega} \frac{\partial}{\partial n} G_k(q(s), q(s'))\, u(s')\,\mathrm{d}s' . \tag{22}$$

For the derivative of the Green function one obtains

$$\frac{\partial}{\partial n} G_k(q(s), q(s')) = \frac{\mathrm{i}k}{4} \cos(\phi(s, s')) H_1^{(1)}(k\,\tau(s, s')) , \tag{23}$$

where $\tau(s, s') = |q(s) - q(s')|$ is the Euclidean distance between the two points on the boundary and

$$\cos\phi(s, s') = \frac{n(s)\,(q(s) - q(s'))}{\tau(s, s')} . \tag{24}$$

This gives the integral equation for the normal derivative $u(s)$

$$u(s) = \oint_{\partial\Omega} Q_k(s, s')\, u(s')\, \mathrm{d}s' \ , \tag{25}$$

with integral kernel

$$Q_k(s, s') = -\frac{\mathrm{i}k}{2} \cos\phi(s, s')\, H_1^{(1)}\left(k\, \tau(s, s')\right) \ . \tag{26}$$

Equation (25) is a Fredholm equation of second kind which has non-trivial solutions if the determinant

$$D(k) := \det(1 - \widehat{Q}_k) \tag{27}$$

vanishes. Here $\widehat{Q}_k$ is the integral operator on $\partial\Omega$ defined by

$$\widehat{Q}_k(u(s)) = \oint_{\partial\Omega} Q_k(s, s')\, u(s')\, \mathrm{d}s' \ . \tag{28}$$

For eigenvalues E_n of the Helmholtz equation with Dirichlet boundary conditions one has $D(k) = 0$ for $k = \sqrt{E_n}$, see e.g. [74] for a detailed proof. However, for $\mathrm{Im}\, k < 0$ there can be further zeros of $D(k)$ which (for the interior Dirichlet problem) correspond to the outside scattering problem with Neumann boundary conditions [76–78] (see also [51]). Explicitly this can be seen from the factorization $D(k) = D(0)D_{\mathrm{int}}(k)D_{\mathrm{ext}}(k)$, where the factors can be written exclusively in terms of the interior and exterior problem. More aspects concerning the additional spurious solutions will be discussed in Sect. 3.3.

Before turning to the numerical implementation, let us discuss the behaviour of the integral kernel for small arguments. The Hankel function $H_1^{(1)}(x)$ reads for small arguments

$$H_1^{(1)}\left(k\, \tau(s, s')\right) \sim -\frac{2\mathrm{i}}{\pi k|s - s'|} \ , \qquad \text{for } s - s' \to 0 \ . \tag{29}$$

This singularity is compensated by the behaviour of

$$\cos\phi(s, s') \sim -\frac{1}{2}\kappa(s)\, |s - s'| \ , \qquad \text{for } s' \to s \ , \tag{30}$$

where $\kappa(s)$ is the curvature of the boundary in the point s. Here the curvature is defined by $\kappa(s) = q_1'(s)q_2''(s) - q_2'(s)q_1''(s)$ such that for example $\kappa(s) = 1$ for a circle of radius one. Thus for the integral kernel we obtain

$$Q_k(s, s') \to \frac{1}{2\pi}\kappa(s) \ , \qquad \text{for } s - s' \to 0 \ . \tag{31}$$

Desymmetrization

For systems with symmetries the numerical effort can be reduced by considering instead of the full system the symmetry reduced system with the corresponding Green function, see e.g. [56]. For a reflection symmetry with respect to the q_1-axis one has

$$G_k^{\pm}(\boldsymbol{q}, \boldsymbol{q}') := G_k(|\boldsymbol{q} - \boldsymbol{q}'|) \pm G_k(|\boldsymbol{q} - (q_1', -q_2')|) \ , \tag{32}$$

where $+$ applies to the case of even eigenfunctions (i.e. Neumann boundary conditions on the symmetry axis) and $-$ to odd eigenfunctions (i.e. Dirichlet boundary conditions on the symmetry axis).

For a two-fold reflection symmetry (as in the case of the stadium billiard, see Fig. 15 for a sketch of the geometry and notations) one has altogether four different subspectra, corresponding to DD, DN, ND and DD boundary conditions on the symmetry axes q_1 and q_2, respectively. For example for Dirichlet-Dirichlet boundary conditions on the q_1- and q_2-axes the Green function reads

$$\begin{aligned} G_k^{\mathrm{DD}} =& G_k(|\boldsymbol{q} - \boldsymbol{q}'|) - G_k(|\boldsymbol{q} - (q_1', -q_2')|) \\ &+ G_k(|\boldsymbol{q} - (-q_1', -q_2')|) - G_k(|\boldsymbol{q} - (-q_1', q_2')|) \ . \end{aligned} \tag{33}$$

For Neuman boundary conditions on these two axes one gets

$$\begin{aligned} G_k^{\mathrm{NN}} =& G_k(|\boldsymbol{q} - \boldsymbol{q}'|) + G_k(|\boldsymbol{q} - (q_1', -q_2')|) \\ &+ G_k(|\boldsymbol{q} - (-q_1', -q_2')|) + G_k(|\boldsymbol{q} - (-q_1', q_2')|) \ . \end{aligned} \tag{34}$$

The advantage of exploiting the symmetries of the system is two-fold: firstly, we can separate the eigenvalues and eigenfunctions for the different symmetry

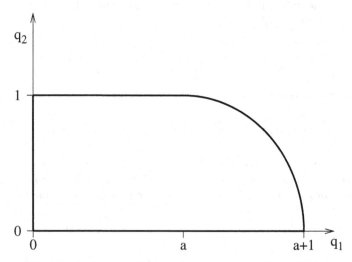

Fig. 15. Geometry of the desymmetrized stadium billiard.

classes, which is necessary for the investigation of the spectral statistics. Secondly, the numerical effort is reduced, since the integral over the whole boundary $\partial\Omega$ is reduced to an integral over a part of the boundary, which in the above examples is half or a quarter of the original boundary. The boundary along the symmetry axes need not be discretized as the boundary condition is already fulfilled by construction. Of course, for other geometries different choices for G can be more appropriate.

Finding the Eigenvalues

In the numerical computations the integral over the boundary is replaced by a Riemann sum. (There also exist more refined methods using polynomial approximations combined with Gauß-Legendre integration, see e.g. [60], which allow for a less fine discretization.) Let $\Delta s = \mathcal{L}/N$ be the discretization length of the boundary of length $\mathcal{L}$ into N pieces. Then we have

$$u(s_i) = \Delta s \sum_{j=0}^{N-1} Q_k(s_i, s_j)\, u(s_j) \; , \qquad (35)$$

where $s_i = (i + 1/2)\Delta s$, $i = 0, \ldots, N - 1$. Equation (35) can be written in matrix form as

$$A_k u = 0 \, , \qquad \text{with} \quad A_{ij} = \delta_{ij} - \Delta s\, Q_k(s_i, s_j) \; . \qquad (36)$$

Recall that for $s_i = s_j$ the kernel $Q_k(s_i, s_j)$ reduces to the result given in (31). The solutions of this linear equation provide approximations to the eigenvalues k_n^2 and eigenvectors u_n. This leads to the problem of finding the real zeroes of the determinant

$$\det(A_k) = 0 \qquad (37)$$

as a function of $k = \sqrt{E}$, where A_k is a dense, complex non-Hermitean matrix. Due to the discretization of the integral the determinant $\det(A_k)$ will not become zero but only close to zero (actually, the discretization shifts the zeros slightly away from the real axis, see [58,59]).

In the numerical computations it is very useful [60] to compute the singular values of the matrix A instead of its determinant. The singular value decomposition of a complex matrix is given by the product of an unitary matrix U, a diagonal matrix S and a second unitary matrix V

$$A = USV^\dagger \; . \qquad (38)$$

The diagonal matrix S contains as entries SV_i the singular values of A and we have $|\det A| = |\prod \mathrm{SV}_i|$. Since the original integral equation has been discretized, the smallest singular value in general never gets zero, but just very small, see Fig. 16. Thus the minima of the smallest singular value provide

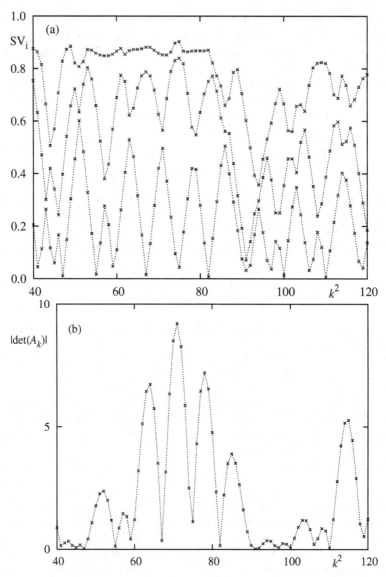

Fig. 16. In a) the three smallest singular values are shown as a function of the energy $E = k^2$ for the stadium billiard with $a = 1.8$ and odd-odd symmetry. The eigenvalues are located at the minima of the first singular value. The second and third singular values allow to locate places with near degeneracies as next to $k^2 = 90$, which can be resolved by magnification of the corresponding region, see Fig. 17. In b) $|\det(A_k)|$ is shown. The minima tend to be not as pronounced as those of the singular values.

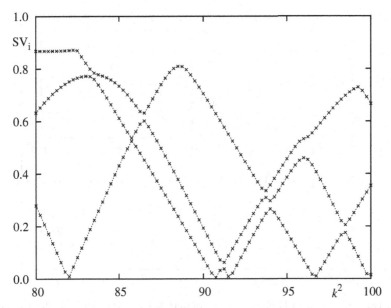

Fig. 17. A magnification of Fig. 16 shows that the singular value decomposition method easily allows to locate nearly degenerate energy levels.

approximations to the eigenvalues of the integral equation. For the numerical computation of the singular value decomposition one may for example use the NAG routine F02XEF or the LAPACK routines ZGESVD or ZGESDD. It turns out that the (more recent) routine ZGESDD is significantly faster (factor 3-5, at the expense of a higher memory consumption), in particular when also singular vectors are computed.

The advantage of the singular value decomposition in comparison to locating the zeros of the determinant is that degeneracies of eigenvalues can be detected by looking at the second singular value, which also gets small when there are two nearby eigenvalues (similarly higher degeneracies can be found by looking at the next singular values). In Fig. 16a) an example of the behaviour of the three smallest singular values is shown in the case of the stadium billiard ($a = 1.8$) with Dirichlet boundary conditions. For comparison a plot of $|\det(A_k)|$ is shown in Fig. 16b). One clearly sees that the singular value decomposition provides more information. For example, next to $k^2 = 90$ the minimum of $|\det(A_k)|$ looks slightly broader than the others, however, this does not give a clear indication that there might be more than one eigenvalue. In contrast, the singular value decomposition method allows to resolve such kind of near-degeneracies efficiently, see Fig. 17. Of course, this information is also available via $\det(A_k)$, see Fig. 18 where its real and imaginary part are plotted separately. Here (approximately) simultaneous zeros correspond to minima of $|\det(A_k)|$. However, notice that compared to

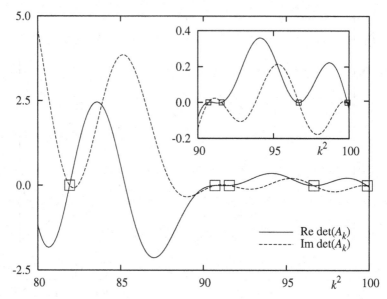

Fig. 18. Plot of real and imaginary part of $\det(A_k)$ as a function of k; the evaluation was done for 10 times as many points in k^2 than for Fig. 17. Approximately simultaneous zeros correspond to minima of $|\det(A_k)|$. The locations of the eigenvalues are marked by squares.

the singular value decomposition approach much more discretization points in $E = k^2$ are necessary.

To determine all energy levels in a given energy interval $[E_1, E_2]$ one proceeds in the following way: first one computes the singular values at equidistantly chosen points $k^2 \in [E_1, E_2]$; the energy is chosen as variable because for two-dimensional billiards the mean distance between two energy levels is approximately constant and according to the generalized Weyl formula (9) given by $\frac{4\pi}{\mathcal{A}}$. The finer the step size is chosen the easier the minima can be resolved, however, at the same time the computing time to cover a given energy range increases correspondingly. The actual step size is a compromise between these two aspects; good results have been achieved by using a step size of the order of $\frac{1}{5}\frac{4\pi}{\mathcal{A}}$ (for systems with many near level degeneracies, e.g. integrable or near-integrable systems, a smaller step size can be helpful).

The matrix size N is chosen according to $N = b\frac{\mathcal{L}}{\lambda} = b\frac{\mathcal{L}k}{2\pi}$, such that one obtains b discretization points per units of the inverse of the de Broglie wave length $\lambda = \frac{2\pi}{k}$ along the boundary $\mathcal{L}$. Typical choices for b are between 5 and 12 depending on the system and the wanted accuracy.

From the first scan one locates all minima of the smallest singular value. If also the second singular value has a minimum next to a minimum of the first one, one has to use a refined discretization in E around the minimum (the numerical implementation is a bit more sophisticated, in order to account for several special situations, so that only a minimal number of additional points need to be computed). Once an isolated minimum is found, an approximation to the eigenvalue can be computed by different methods. Either one can perform a refined computation around the minimum, which can be quite time-consuming, or one can use a local approximation by a parabola [79]. A linear interpolation also gives good results: From the three points $1:(k_1^2, SV_1(k_1^2))$, $2:(k_2^2, SV_1(k_2^2))$, $3:(k_3^2, SV_1(k_3^2))$, characterizing a minimum of the first singular value, one has two different lines $\overline{12}$ and $\overline{23}$ with different slopes, of which the line with the larger slope has to be chosen. The intersection of this line with the zero axis gives a good approximation to the eigenvalue, which one can refine if necessary. By repeatedly applying this for all minima, all energy levels in a given interval can be found. In fact, it is possible to develop a computer program which takes care of all this such that all levels can be found automatically.

A good check of the completeness is provided by considering the δ_n statistics, see the example in Sect. 3.2. The accuracy of the computed eigenvalues can be estimated from the bracket of the minimum given by the three points 1,2,3 if the matrix dimension N is large enough. For N too small (for a given resolution in E) one does not obtain a peaked, but a broad minimum. This is illustrated in Fig. 19 by magnifying Fig. 17 around the minimum with $k^2 \approx 96.5$ for different N. One clearly sees the parabolic structure around

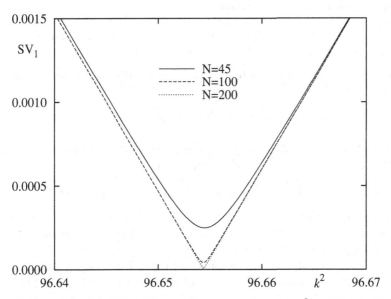

Fig. 19. Magnification of Fig. 17 around the minimum with $k^2 \approx 96.5$ for different matrix sizes N. One nicely sees the pronounced parabolic structure for $N = 45$ which gets smaller for larger N.

the minimum for smaller N and for larger N one recovers the essentially linear behaviour of the smallest singular value.

Tests of the accuracy of the method can be obtained by considering a system where the eigenvalues are known. For example for the circular billiard the eigenvalues can be computed with arbitrary accuracy. Also billiards where the eigenvalues can be computed by other methods (e.g. conformal mapping method [80, 81]) allow a determination of the accuracy of the method. For a study of the scaling of the error for various billiards see [63]. In addition computations of the normal derivative function $u_n(s)$ and the eigenfunction (both inside and outside of Ω) allow to check the quality of the numerical method and program.

Computing Eigenfunctions

From a minimum of the smallest singular value we obtain an approximation of the eigenvalue and at the same time the corresponding singular vector u gives an approximation to the normal derivative function $u(s)$. The NAG routine F02XEF scales the singular vector such that its first component is real. Thus for a correct solution also the other components should be essentially real, which provides another check for the implementation of the method and the accuracy of the eigenvalues.

The eigenfunction in the interior of the domain Ω can now be calculated from the normal derivative function,

$$\psi(\boldsymbol{q}) = -\frac{i}{4} \oint_{\partial\Omega} H_0^{(1)}\left(k\left|\boldsymbol{q} - \boldsymbol{q}(s)\right|\right) u(s)\,\mathrm{d}s , \qquad \text{for } \boldsymbol{q} \in \Omega\backslash\partial\Omega . \qquad (39)$$

The computation of the eigenfunction can be simplified by taking into account that

$$\oint_{\partial\Omega} J_0\left(k\left|\boldsymbol{q} - \boldsymbol{q}(s)\right|\right) u(s)\,\mathrm{d}s = 0 , \qquad (40)$$

because the J_0-part of $G_k(\boldsymbol{q}, \boldsymbol{q}')$ is a solution of the homogeneous equation corresponding to (16). Thus (39) is equivalent to

$$\psi(\boldsymbol{q}) = \frac{1}{4} \oint_{\partial\Omega} Y_0\left(k\left|\boldsymbol{q} - \boldsymbol{q}(s)\right|\right) u(s)\,\mathrm{d}s , \qquad \text{for } \boldsymbol{q} \in \Omega\backslash\partial\Omega . \qquad (41)$$

If one uses a desymmetrization, such as (32), (33) or (34), the above formula (41) has to be modified accordingly.

In Fig. 20 we show some examples of normal derivatives $u_n(s)$ and the corresponding eigenfunctions of the billiard, computed via (41). The imaginary part of $u_n(s)$ is typically 5 or more orders of magnitude smaller than the real part. It is interesting to see that part of the structure of the eigenfunctions is also reflected in $u_n(s)$. For example for eigenstates with small probability in the region of the quarter circle also the normal derivative is small for $s < \pi/2$.

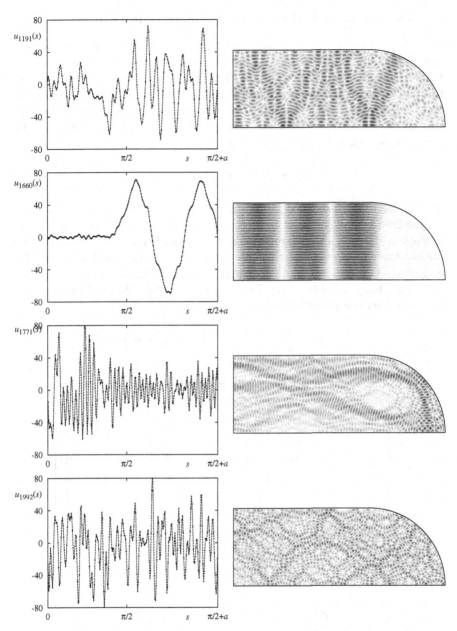

Fig. 20. Examples of normal derivative functions $u_n(s)$ and the corresponding eigenfunctions in the stadium billiard (odd-odd symmetry, $a = 1.8$). Here black corresponds to high intensity of $|\psi_n(\boldsymbol{q})|^2$.

Spurious Solutions I: Real Green Function Approach

In certain situations and for some numerical methods it may happen that one obtains in addition to the true solutions of the Helmholtz equation (14) further so-called *spurious solutions*. This question is discussed in some of the papers on the boundary integral method, in particular see [50, 52, 53] and [55, 58]. There are essentially two different situations in which they are encountered. The first is that one uses for the Green function instead of the Hankel function, see (21), just the real part, i.e.

$$G_k(\boldsymbol{q}, \boldsymbol{q}') = \frac{1}{4} Y_0 \left(k \left| \boldsymbol{q} - \boldsymbol{q}' \right| \right) \ . \tag{42}$$

This seems reasonable as according to (40) the J_0-Bessel function does not contribute to the eigenfunction. Moreover then one can work with an entirely real matrix for which the singular value decomposition can be computed much faster. However, when using this approach, there appear additional zeros (for each correct one there is one additional one) and the singular values loose their nice linear structure, see Fig. 21. To overcome the problem of these additional zeros a parametrized Green function

$$G_k^{(\beta)}(\boldsymbol{q}, \boldsymbol{q}') = \frac{1}{4} \left[\beta J_0 \left(k \left| \boldsymbol{q} - \boldsymbol{q}' \right| \right) + Y_0 \left(k \left| \boldsymbol{q} - \boldsymbol{q}' \right| \right) \right] \tag{43}$$

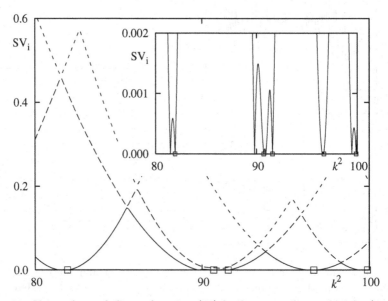

Fig. 21. Using the real Green function (42) leads to spurious solutions (see the inset) in addition to the correct eigenvalues marked by squares (compare with Fig. 17). For each true solution there is an additional spurious one (hardly visible at $k^2 \approx 91$ and $k^2 \approx 96$).

is used in [58]. Thus for $\beta = 0$ we obtain (42) and for $\beta = -i$ we get (21).
So using a purely real Green function means to vary $\beta \in \mathbb{R}$ which changes
the location of the spurious solutions but not those of the true ones. This is
illustrated in Fig. 22 around the eigenvalue $k^2 = 81.93\ldots$ with $\beta \in [0, 0.1]$.
Clearly on this scale the true solution does not change under variation of β
(apart from the region of the avoided crossing which is due to the finite matrix
size and gets smaller for larger N) whereas the spurious solution strongly
varies with β. For $\beta = -\gamma i$ with increasing real γ the additional zeros move
away from the real axis and it seems that for $\beta = -i$ they do not have any
significant influence on the real axis. Still there could be cases where also for
$\beta = -i$ such a solution becomes relevant, but for convex geometries we have
not encountered this situation. For an example of a non-convex geometry see
Sect. 3.3.

As an explicit example for the influence of parameterized Green function
(43) let us consider the circular billiard with radius 1, where the Fredholm
determinant reads (see e.g. [58, 74])

$$D(k) = \prod_{l=-\infty}^{\infty} \left[-i\pi k H_l^{(1)'}(k) J_l(k) \right] . \tag{44}$$

As this product converges absolutely in the whole complex k-plane (apart
from a cut along the negative real axis) zeros of $D(k)$ occur when one of the

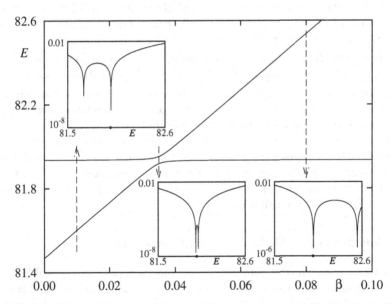

Fig. 22. Plot of the minima of the singular values around the eigenvalue $k^2 = 81.93\ldots$ with varying β using the parametrized Green function (43). The insets show the corresponding structure of the first singular value with a logarithmic vertical scale (matrix size for this computation: $N = 200$).

factors in the product vanishes [74]. Clearly, the real zeros of $D(k)$ correspond to the eigenvalues j_{ml} of the circular billiard with radius 1 and Dirichlet boundary conditions. The further zeros stem from the functions $H_l^{(1)'}(z)$ which have only zeros with $\mathrm{Im}\, z < 0$ [82], and do not correspond to physical solutions of the interior problem. However, they can be related to resonances of the exterior scattering problem, but with Neumann boundary conditions [76,77]. Because of the radial symmetry the S-matrix is diagonal in angular momentum space

$$S_{l'l} = -\frac{H_l^{(2)'}(k)}{H_l^{(1)'}(k)}\delta_{l'l} \tag{45}$$

and therefore the resonances are at those complex k for which

$$H_l^{(1)'}(k) = 0 \ , \tag{46}$$

i.e. the same condition as implied by (44).

If one uses the parametrized Green function (43) one can show (analogous to the derivation of (44)) that for the circular billiard

$$D^{(\beta)}(k) = \prod_{l=-\infty}^{\infty} \left[\pi k \left(\beta J_l'(k) + Y_l'(k)\right) J_l(k)\right] \ . \tag{47}$$

For $\beta = 0$, which corresponds to the real Green function (42), we get additional zeros of $D^{(0)}(k)$ when $Y_l'(k) = 0$. Varying β from zero to $-i$ these real zeros turn complex. At first sight one might think that these are connected to the places with $H_l^{(1)'}(k) = 0$, however numerical computations show that (for all studied cases) these move away from the real axis with a positive imaginary part and for $\beta = -i$ one has $H_l^{(1)'}(k) = 0$ only for $\mathrm{Im}\, k < 0$. Thus the spurious solutions for the real Green function are not related to resonances of the scattering problem with Neumann boundary conditions.

These examples suggest to use the full complex Green function (21) instead of the real variant (42). Even though the numerical computation is more time-consuming for the complex case their advantages over choosing (42) are obvious as the variation of β is time-consuming as well (and non-trivial to implement in an automatic way).

Spurious Solutions II: Non-convex Geometries

Even when choosing the complex Green function (21) it is possible to encounter spurious solutions: For the circular the additional complex zeros of $D(k)$ are sufficiently far away from the real axis, i.e. $\mathrm{Im}\, k \ll 0$ so that they do not lead to problems with the application of the boundary integral method. However, when one considers different geometries the resonances of the corresponding scattering system could be closer to the real axis. This can be

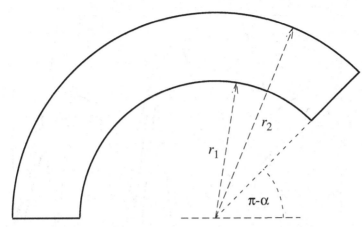

Fig. 23. Boundary of the annular sector billiard for $\alpha = \frac{7}{8}\pi$ and $r_1 = 0.4$ and $r_2 = 0.6$.

nicely studied for the annular sector billiard, see Fig. 23, as the eigenvalues and eigenfunctions can be determined numerically with arbitrary accuracy. Using the ansatz [83, §25]

$$\psi(r,\phi) = [J_\nu(kr) + cY_\nu(kr)]\sin(\nu\phi) \tag{48}$$

with $\nu = m\frac{\pi}{\alpha}$, $m = 0, 1, 2, \ldots$ and requiring $\psi(r_1, \phi) = 0$ and $\psi(r_2, \phi) = 0$ gives the (implicit) eigenvalue equation

$$J_\nu(kr_1)Y_\nu(kr_2) - Y_\nu(kr_1)J_\nu(kr_2) = 0 \ . \tag{49}$$

For each m one gets a sequence of zeros $k_{mn} = \sqrt{E_{mn}}$.

Figure 24 shows for the annular sector billiard with $\alpha = \frac{49}{50}\pi$ the first three singular values as a function of k^2. The solutions of (49) are marked by triangles. Clearly, there are additional minima, which can be associated with resonances of the dual scattering problem (for further details and examples of this association for the annular sector billiard see [84]). In the limit of $\alpha \to \pi$ these resonances are given by the eigenvalues of the circular billiard of radius r_1 with Neumann boundary conditions. For this billiard the ansatz $\psi(r,\phi) = J_m(kr)$ together with $\left.\frac{\partial\psi(r,\phi)}{\partial r}\right|_{r=r_1} = 0$ gives the eigenvalue equation

$$mJ_m(kr_1) - kr_1 J_{m+1}(kr_1) = 0 \ . \tag{50}$$

The circles shown in Fig. 24 correspond to the solutions of (50) and provide a very good description of the additional minima.

Thus the question arises how to detect and distinguish these additional solutions. First, of course their existence and relevance strongly depends on the system one is studying. In many situations (for example convex geometries) there appear to be no complex solutions coming close enough to the

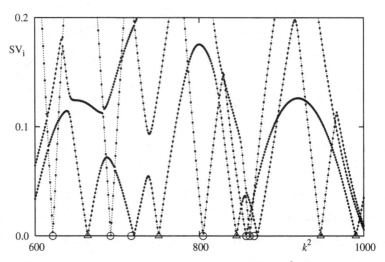

Fig. 24. First three singular values as a function of $E = k^2$ of the annular sector billiard for $\alpha = \frac{49}{50}\pi$ and $r_1 = 0.4$ and $r_2 = 0.6$. The triangles correspond to the exact eigenvalues for the annular sector billiard, computed from (49) and the circles correspond to the eigenvalues of the circular billiard with radius r_1 and Neumann boundary condition, determined via (50).

real axis. Intuitively this seems reasonable as long as there are no trapped orbits outside of the billiard as these should give rise to resonances with small imaginary part.

However, if such additional solutions exist they will show up in the δ_n statistics by an offset of $+1$ at each additional eigenvalue (unless one by chance misses the same number of 'correct' eigenvalues). If one has a system with such additional solutions one approach is to plot the corresponding normal derivative function $u(s)$ and the eigenfunction. Usually they will behave quite differently for a correct eigenvalue and for a spurious solution. For example for the case of the annular sector billiard the normal derivative function for a spurious solution is discontinuous along the boundary and the corresponding eigenfunction also has contributions outside of the billiard, see Fig. 26. Another test would be to use the normalization condition (51) for the normal derivative and compute the norm of the eigenfunction in the interior of the billiard. These two are the same for proper eigenfunctions whereas for spurious solutions they will disagree. Unfortunately, this is a highly inefficient method as the computation of the eigenfunction in Ω is quite time-consuming. Instead of computing the normalization for the full billiard one could restrict to smaller subregions, e.g. for the annular sector billiard one could integrate over the region of the circle with radius r_1 and check if it is different from zero indicating a spurious solution. For the annular sector billiard the additional zeros of the Fredholm determinant $D(k)$ are complex as long as $\alpha < \pi$. Thus for $N \to \infty$ these minima will stay bounded away from zero in contrast to the minima corresponding to the eigenvalues. However, in practice it is not possible to check this as one has to make N too large to distinguish these from the correct eigenvalues.

More generally, spurious solutions can be understood by a second look at the boundary integral equations. Namely, for the interior Dirichlet problem we have the *single layer* equation, (19), and the *double layer* equation, (25). On the other hand, the single-layer equation for the outside scattering problem with Neumann boundary conditions at $\partial\Omega$ is also given by the double layer equation (25). (see e.g. [50, 51]). As a consequence, scattering solutions of the outside scattering problem with Neumann boundary conditions at $\partial\Omega$ may become relevant for real k. Namely, for resonances with small imaginary part they can lead to additional solutions for the double layer equation which are numerically indistinguishable from the correct solutions. However, these solutions do not correspond to solutions of the interior problem and they do not fulfill the single layer equation. So a possibility to distinguish spurious solutions for the interior Dirichlet problem is to check the validity of the single layer equation as well, which is only fulfilled simultaneously for correct solutions of the interior Dirichlet problem.

A common approach (see e.g. [50] and references therein) to incorporate this from the beginning is by combining the single layer and double layer equation using a linear superposition. By this the solutions of the outside problem with Neumann boundary conditions can be removed. Because of the singular kernel in the single layer equation special care has to be taken with

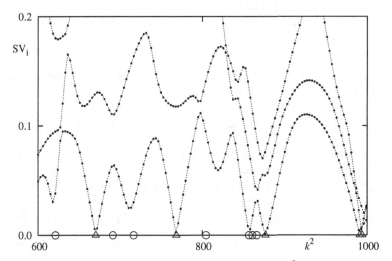

Fig. 25. First three singular values as a function of $E = k^2$ of the annular sector billiard for $\alpha = \frac{7}{8}\pi$ and $r_1 = 0.4$ and $r_2 = 0.6$. The triangles correspond to the exact eigenvalues for the annular sector billiard, computed from (49) and the circles correspond to the eigenvalues of the circular billiard with radius r_1 and Neumann boundary condition, determined via (50)

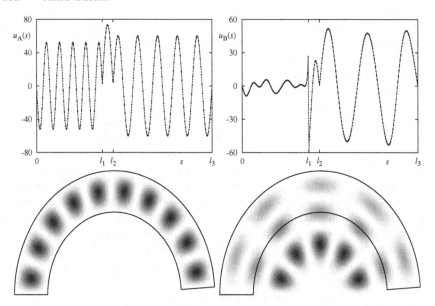

Fig. 26. Normal derivative functions $u_n(s)$ corresponding to the correct eigenvalue with $E = 663.88\dots$ (left) and the spurious one with $E = 691.77\dots$ (right). Here $l_1 = r_1\alpha$, $l_2 = r_1\alpha + r_2 - r_1$ and $l_3 = r_1\alpha + r_2 - r_1 + \alpha r_2$. One clearly sees the discontinuity in $u_n(s)$ for the spurious solution. This is also reflected in the structure of the eigenfunction which for the spurious solution has its main contribution outside of the billiard. Notice that in both cases the eigenfunction has been computed according to (40) inside and outside of Ω. The fact that for the correct eigenfunction $\psi_n(q) = 0$ (within the numerical accuracy) for $q \in \mathbb{R}^2 \backslash \Omega$ is another test of the accuracy of the eigenvalue computations and eigenfunctions.

the implementation. For the more difficult case of billiards with magnetic field see [65].

Derived Quantities in Terms of the Normal Derivative Function

As the normal derivative function contains all information to determine the eigenfunction, it is interesting to see if this approach can be used to compute other quantities of interest. For example, if one wants to calculate expectation values $\langle \psi | A | \psi \rangle$ of some operator A in the state ψ, one has to ensure that the eigenfunction ψ is normalized, $\langle \psi | \psi \rangle = \int_\Omega |\psi(q)|^2 \, \mathrm{d}^2 q = 1$. In principle this could be done by considering $\left(\langle \widetilde{\psi} | \widetilde{\psi} \rangle \right)^{-1} \widetilde{\psi}(q)$ of an unnormalized eigenfunction $\widetilde{\psi}$. However, an accurate computation of $\langle \widetilde{\psi} | \widetilde{\psi} \rangle$ using (41) is quite time consuming. Fortunately, there is a simpler way to achieve a normalized ψ: If ψ is a normalized eigenfunction with eigenvalue $E = k^2$ and $u(s)$ is the corresponding normal derivative then we have the following normalization condition for $u(s)$ [55, 59]

$$\frac{1}{2} \oint_{\partial\Omega} \boldsymbol{n}(s)\boldsymbol{q}(s)\,|u(s)|^2\,\mathrm{d}s = k^2 \ . \tag{51}$$

If $\widetilde{u}(s)$ is an unnormalized normal derivative, then one obtains by

$$u(s) = \frac{\sqrt{2}\,k}{\sqrt{\oint_{\partial\Omega} \boldsymbol{n}(s)\boldsymbol{q}(s)\,|\widetilde{u}(s)|^2\,\mathrm{d}s}}\,\widetilde{u}(s) \tag{52}$$

a normalized one. Starting with a normal derivative normalized in this way, any other quantities (e.g. expectation values) determined in terms of $u(s)$ have the correct normalization.

This is just the first example out of many highlighting the importance of the normal derivative for numerical computations of quantities related to eigenfunctions. For example, there are explicit expressions in terms of $u_n(s)$ to compute the

– normalization of ψ, (51), [55, 59]
– eigenfunction ψ, (41)
– momentum distribution

$$\widehat{\psi}_n(\boldsymbol{p}) = \frac{1}{2\pi}\int_\Omega \mathrm{e}^{-\mathrm{i}\boldsymbol{p}\boldsymbol{q}}\psi_n(\boldsymbol{q})\,\mathrm{d}^2 q = -\frac{\mathrm{i}}{4\pi p_n^2}\int_{\partial\Omega}\mathrm{e}^{-\mathrm{i}\boldsymbol{p}\boldsymbol{q}(s)}\boldsymbol{p}\boldsymbol{q}(s)u_n(s)\,\mathrm{d}s\,, \tag{53}$$

and radially integrated momentum distribution [85, 86]

$$I(\varphi) := \int_0^\infty \left|\widehat{\psi}_n(r,\varphi)\right|^2 r\,\mathrm{d}r \ , \tag{54}$$

see [86] for details.
– Husimi functions (see e.g. [87, 88])
– autocorrelation function of eigenstates [89].

In Figs. 27-29 we show for the cardioid billiard examples of eigenfunctions in position space, the corresponding momentum distributions, the angular momentum distributions (for further details and examples see [86]) and the corresponding Husimi functions $H_n(s,p)$. The first example in Fig. 27 shows an example of a scarred state, i.e. an eigenstate which shows localization round an unstable periodic orbit [90]. Below the three-dimensional plot of the state is the corresponding density plot (black corresponding to high intensity) in which the localization is clearly visible. Also the corresponding three-dimensional plot of the momentum distribution $\widehat{\psi}_{567}(\boldsymbol{p})$ reveals enhanced contributions in the directions $\varphi = \pi/2, 3\pi/2$. This is also seen in the plot of $I_{567}(\varphi)$ which shows that the probability to find the particle with momentum near $\pi/2$ is significantly enhanced compared to the mean of $1/(2\pi)$.

$n = 567$, odd symmetry

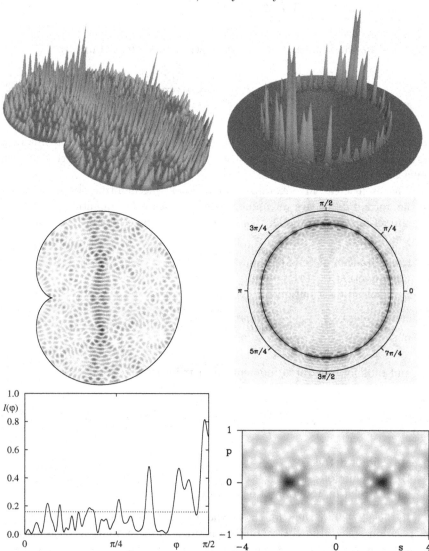

Fig. 27. Three-dimensional plots of $|\psi_{567}(\boldsymbol{q})|^2$, $|\widehat{\psi}_{567}(\boldsymbol{p})|^2$, their corresponding grey-scale pictures and the plot of the radially integrated momentum distribution $I_{567}(\varphi)$. The momentum distribution $|\widehat{\psi}_{567}(\boldsymbol{p})|^2$ is concentrated around the energy shell, which is indicated by the inner circle. This state is localized along the shortest unstable periodic orbit, leading to an enhancement of $|\widehat{\psi}_{567}(\boldsymbol{p})|^2$ near to $\varphi = \pi/2, 3\pi/2$, also seen in the plot of $I_{567}(\varphi)$ near to the momentum direction $\varphi = \pi/2$ (marked by a triangle). This localization is also clearly visible in the Husimi representation.

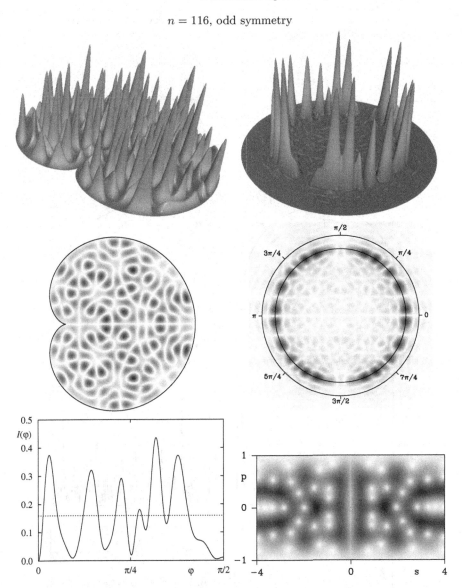

$n = 116$, odd symmetry

Fig. 28. Same as in the previous figure but for $n = 116$. In this case there is no prominent localization neither in position nor in momentum space.

Another representation is the Husimi-Poincaré representation $H_n(s, p)$ where s corresponds to the arclength coordinate along the billiard boundary and p corresponds to the projection of the unit velocity vector after the reflection on the tangent in the point s. In this picture the localization around the unstable orbit is maybe most clearly seen; the places of high intensity are on the

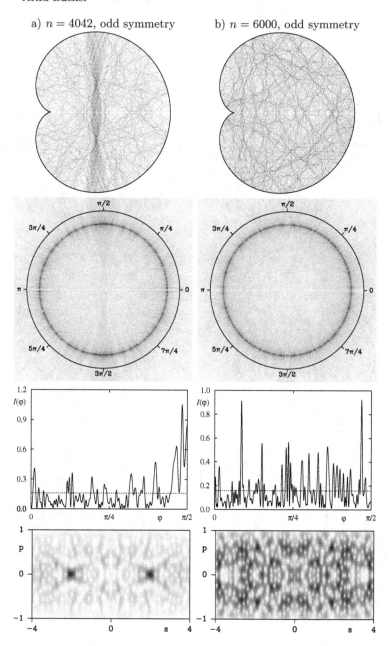

Fig. 29. The eigenfunction in a) shows localization along the shortest unstable orbit which is also reflected in the momentum distributions and in the Husimi function. The eigenfunction in b) is an example which appears to be quite delocalized both in position and in momentum space. The pictures look like those expected (according to the quantum ergodicity theorem) for a 'typical' eigenfunction.

line $p = 0$ (perpendicular reflection) and match perfectly with the position of the orbit.

The second example shown in Fig. 28 is an 'ergodic' state, i.e. a state which does not show any significant localization (as much as something like this is possible at low energies) neither in position nor in momentum space (apart from the localization on the energy shell). This is nicely reflected in the various representations. Two further examples are shown in Fig. 29 where a) is a higher lying scar and b) is another 'typical' state (in the sense of the quantum ergodicity theorem).

4 Concluding Remarks – Or What's Left ?

There are many more issues related to scientific computing in quantum chaos which I did not mention in these notes. They for example include visualization techniques, programming of parallel computers (e.g. using PVM or MPI), or using vector computers etc. Also the more implementation specific aspects, including the choice of a programming language have not been discussed. A good starting point to learn about computing in quantum chaos are quantum maps as their numerics is much easier (one can use a black-box routine to get all eigenvalues at once) than for billiard systems, where more complicated methods have to be used.

Acknowledgements

I would like to thank the organizers of the summer school *The Mathematical aspects of Quantum Chaos I* in Bologna, Mirko Degli Esposti and Sandro Graffi for all their effort, especially Mirko for dealing with everything (including food and wine ;-) in the 'Italian way'! Moreover, I am grateful to Grischa Haag for many discussions on quantum maps, and Ralf Aurich for discussions on the boundary element method. I would like to thank Professor Frank Steiner and Silke Fürstberger for useful comments on the manuscript and Professors Uzy Smilansky and Andreas Knauf for useful remarks on an earlier version of this text. I would like to thank Fernando Perez for pointing out some speed improvements of the Python implementation. Most parts of this work have been done during my stay at the School of Mathematics, University of Bristol, and the Basic Research Institute in the Mathematical Sciences, Hewlett-Packard Laboratories Bristol, UK. In particular I would to thank Professors Jonathan Keating and Sir Michael Berry for all their support. Moreover, I am particularly indebted to Professor Steiner and the University of Ulm for their support throughout the last months. I acknowledge partial support by the Deutsche Forschungsgemeinschaft under contract No. DFG-Ba 1973/1-1.

Appendix: Computing Eigenvalues of Quantum Maps

The first thing when thinking of solving a certain problem numerically is to decide on the programming language. There are numerous possibilities, ranging from Assembler, Fortran, Pascal, C, C++, Java, etc. to using packages like Octave, Matlab, Maple or Mathematica. Here I will use the quite recent scripting language Python [91]. Of course it is beyond the scope of this text to give an introduction to this language; several excellent introductions can be found on the Python homepage. In addition to the basic Python installation you will also need the Numeric package [92], which is also simple to install. The following programs together with further information can be obtained from [20]. If you have been wondering about the name - yes it originates from Monty Python's flying circus, and at several places the documentation refers to more or less famous Monty Python sketches.

So here is pert_cat.py (the full version can be obtained via [20]):

```python
#!/usr/bin/env python

import cmath
from Numeric import zeros,Float,Complex
from math import sin,pi,sqrt
import LinearAlgebra

def quantum_cat(N,kappa):
    """For a given N and kappa this functions returns the
    corresponding unitary matrix U of the
    quantized perturbed cat map.
    """
    mat=zeros((N,N), Complex) # complex matrix with NxN elements
    I=1j # predefine sqrt(-1)
    # now fill each matrix element
    # (note: this can be done much faster, see the on-line version)
    for k in range(0,N):
    for l in range(0,N):
    mat[k,l]=cmath.exp(2.0*I*pi/N*(k*k-k*l+l*l)+ \
    I*kappa*N/2.0/pi*sin(2.0*pi/N*l))/sqrt(N)
    return(mat)


def compute_evals_pcat(N,kappa):
    """For a given N and kappa this functions returns
    the eigenvalues and eigenphase of the unitary matrix U
    filled via quantum_cat(N,kappa).
    """
    matU=quantum_cat(N,kappa) # fill matrix U_N
    # determine eigenvalues of U_N:
    evals=LinearAlgebra.eigenvalues(matU)

    # determine phase \in [0,2\pi] from the eigenvalues
```

```
phases_N = arctan2(evals.imag,evals.real) + pi
# useful to determine level-spacing
phases = concatenate([phases_N,[phases_N[0]+2.0*pi]])
return(evals,phases)

### Main (used if pert_cat.py is called as script)
if __name__ == '__main__':
from string import atoi,atof
import sys

# Determine eigenvalues and eigenphases
(evals,phases)=compute_evals_pcat(atoi(sys.argv[1]), \
atof(sys.argv[2]))

for k in range(0,N): # print eigenvalues
print("% e % e % e % e ") % \
(evals[k].real,evals[k].imag,phases[k],abs(evals[k]))
```

The only drawback of the above code is that the loop to fill the matrix is slower than a corresponding code in C or Fortran (notice that there are some very nice ways of overcoming this by inlining of code or on-the-fly compilation which are presently being developed for example in the context of SciPy [93]). However, as **diagonalize** uses the LAPACK library the most time-consuming part (at least for larger N) is done in an efficient way (not taking into account the possibility of using ATLAS [24] for further speed improvements).

As a first test do (for $N = 101$ and $\kappa = 0.3$)

```
python pert_cat.py 101 0.3
```

It will output the (complex) eigenvalues as a sequence x, y pairs. As a test, whether these all lie on the unit circle the third column is the absolute value of the eigenvectors. To plot the resulting data you may use

```
python pert_cat.py 101 0.3 > pcat_101_0.3.dat
```

which redirects the output of the program to the file pcat_101_0.3.dat. To plot the resulting file use your favourite plotting program, e.g. for **gnuplot** [94] just do

```
plot "pcat_101_0.3.dat" using 1:2 with points
```

Now we would like to compute the level spacing distribution. To do this let us use an interactive Python session in which we do

```
from Numeric import * # Numeric package
from pert_cat import compute_evals_pcat # above pert_cat routines
from AnalyseData import * # histogram (see below)

N=53
kappa=0.3
```

```
(evals,phases)=pert_cat.compute_evals_pcat(N,kappa);
# sort and unfold phases
s_phases=Numeric.sort(phases)*N/(2.0*pi)

# determine Level spacing
# (by computing the difference of the shifted eigenphases)
spacings=s_phases[1:]-s_phases[0:N]

(x_histogram,y_histogram)=histogram(spacings,0.0,10.0,100)
store_histogram(x_histogram,y_histogram,"histogram.dat")
```

Then use your favourite plotting program to plot the level spacing distribution. For gnuplot you could do

```
goe_approx(x)=pi/2.0*x*exp(-pi/4*x*x)
gue_approx(x)=32/pi/pi*x*x*exp(-4/pi*x*x)
plot "histogram.dat" w l,goe_approx(x),gue_approx(x),exp(-x)
```

Here the routines to compute and store the histogram are in AnalyseData.py whose core reads

```
def histogram(data,min,max,nbins):
from Numeric import *
# first select only those which lie in the interval [min,max]
hdat=compress( ((data<max)*(data>min)),data)
bin_width=(max-min)/nbins
# define the bins
bins=min+bin_width*arange(nbins)
# determine indices
inds=searchsorted(sort(hdat),bins)
inds=concatenate([inds,[len(hdat)]])
# return bins and normalized histogram
return(bins,(inds[1:]-inds[:-1])/(bin_width*len(hdat)))

def store_histogram(x_distrib,y_distrib,outdat):
bin_width=x_distrib[1]-x_distrib[0]
f=open(outdat,"w") # open file for writing
for k in range(0,len(x_distrib)):
f.write("% e % e \n" % (x_distrib[k],y_distrib[k]))
f.write("% e % e \n" % (x_distrib[k]+bin_width, \
y_distrib[k]))
f.close()
```

Again, for further details and full routines see [20].

References

1. J. Meiss: *Symplectic maps, variational principles, and transport*, Rev. Mod. Phys. **64** (1992) 795-848.
2. J.-M. Strelcyn: *The "coexistence problem" for conservative dynamical systems: a review*, Colloquium mathematicum **62** (1991) 331–345.

3. P. Duarte: *Plenty of elliptic islands for the standard family of area preserving maps*, Ann. Inst. H. Poincaré Anal. Non Linéaire **11** (1994) 359–409.

4. A. Giorgilli and V. F. Lazutkin: *Some remarks on the problem of ergodicity of the standard map*, Phys. Lett. A **272** (2000) 359–367.

5. V. F. Lazutkin: *A remark on "Some remarks on the problem of ergodicity of the standard map"*, preprint, mp-arc 00-159 (2000).

6. M. Basilio de Matos and A. M. Ozorio de Almeida: *Quantization of Anosov maps*, Annals of Physics **237** (1993) 46–65.

7. P. A. Boasman and J. P. Keating: *Semiclassical asymptotics of perturbed cat maps*, Proc. R. Soc. London Ser. A **449** (1995) 629–653.

8. V. I. Arnold and A. Avez: *Ergodic Problems of Classical Mechanics*, Benjamin, NewYork, (1968).

9. M. V. Berry, N. L. Balazs, M. Tabor and A. Voros: *Quantum maps*, Annals of Physics **122** (1979) 26–63.

10. J. H. Hannay and M. V. Berry: *Quantization of linear maps on a torus — Fresnel diffraction by periodic grating*, Physica D **1** (1980) 267–290.

11. N. L. Balazs and A. Voros: *The quantized Baker's transformation*, Ann. Phys. **190** (1989) 1–31.

12. M. Saraceno: *Classical structures in the quantized baker transformation*, Ann. Phys. **199** (1990) 37–60.

13. M. Degli Esposti: *Quantization of the orientation preserving automorphisms of the torus*, Ann. Inst. H. Poincaré Phys. Théor. **58** (1993) 3 323–341.

14. M. Degli Esposti, S. Graffi and S. Isola: *Classical limit of the quantized hyperbolic toral automorphisms*, Commun. Math. Phys. **167** (1995) 471–507.

15. S. De Bièvre, M. Degli Esposti and R. Giachetti: *Quantization of a class of piecewise affine transformations on the torus.*, Commun. Math. Phys. **176** (1996) 73–94.

16. S. Zelditch: *Index and dynamics of quantized contact transformations.*, Ann. Inst. Fourier **47** (1997) 305–363.

17. G. Haag: *Quantisierte chaotische Abbildungen*, Diploma Thesis, Abteilung Theoretische Physik, Universität Ulm (1999).

18. S. De Bièvre: *Quantum chaos: a brief first visit* in: *Second Summer School in Analysis and Mathematical Physics: Topics in Analysis: Harmonic, Complex, Nonlinear and Quantization*, S. Perez-Esteva and C. Villegas-Blas (eds.), Contemporary Mathematics **289** (2001).

19. T. Prosen and M. Robnik: *Numerical demonstration of the Berry-Robnik level spacing distribution*, J. Phys. A **27** (1994) L459–L466, corrigendum: ibid. **27** (1994) 6633–6633.

20. See http://www.physik.tu-dresden.de/ baecker/qmaps.html for programs and further information.

21. The Numerical Algorithms Group (NAG), http://www.nag.co.uk/.

22. LAPACK – Linear Algebra PACKage, http://www.netlib.org/lapack/.

23. R. Ketzmerick, K. Kruse and T. Geisel: *Efficient diagonalization of kicked quantum systems*, Physica D **131** (1999) 247–253.

24. ATLAS, http://math-atlas.sourceforge.net/ (Automatically Tuned Linear Algebra Software).

25. O. Bohigas, M.-J. Giannoni and C. Schmit: *Characterization of chaotic quantum spectra and universality of level fluctuation laws*, Phys. Rev. Lett. **52** (1984) 1–4.

26. M. V. Berry and M. Tabor: *Level clustering in the regular spectrum*, Proc. R. Soc. London Ser. A **356** (1977) 375–394.

27. J. P. Keating: *Asymptotic properties of the periodic orbits of the cat maps*, Nonlinearity **4** (1991) 277–307.

28. J. P. Keating: *The cat maps: Quantum mechanics and classical motion*, Nonlinearity **4** (1991) 309–341.

29. F. Mezzadri: *Boundary conditions for torus maps and spectral statistics*, Ph.D. thesis, School of Mathematics, University of Bristol, (1999).

30. J. P. Keating and F. Mezzadri: *Pseudo-symmetries of Anosov maps and spectral statistics*, Nonlinearity **13** (2000) 747–775.

31. T. A. Brody, J. Flores, J. B. French, P. A. Mello, A. Pandey and S. S. M. Wong: *Random-matrix physics: spectrum and strength fluctuations*, Rev. Mod. Phys. **53** (1981) 385–479.

32. F. Haake: *Quantum Signatures of Chaos*, Springer-Verlag, Berlin, 2nd edn., (2001).

33. C. E. Porter and R. G. Thomas: *Fluctuations of Nuclear Reaction Widths*, Phys. Rev. **104** (1956) 483–491.

34. P. Kurlberg and Z. Rudnick: *Value distribution for eigenfunctions of desymmetrized quantum maps*, Internat. Math. Res. Notices (2001) 995–1002.

35. B. Eckhardt: *Exact eigenfunctions for a quantised map*, J. Phys. A **19** (1986) 10 1823–1831.

36. A. Bouzouina and S. De Biévre: *Equipartition of the eigenfunctions of quantized ergodic maps on the torus*, Commun. Math. Phys. **178** (1996) 83–105.

37. S. De Bièvre and M. Degli Esposti: *Egorov theorems and equidistribution of eigenfunctions for the quantized sawtooth and baker maps*, Ann. Inst. Henri Poincaré, Physique Théorique **69** (1996) 1–30.

38. A. Bäcker and H. R. Dullin: *Symbolic dynamics and periodic orbits for the cardioid billiard*, J. Phys. A **30** (1997) 1991–2020.

39. H. P. Baltes and E. R. Hilf: *Spectra of Finite Systems*, Bibliographisches Institut, Mannheim, Wien, Zürich, (1976).

40. M. Sieber, U. Smilansky, S. C. Creagh and R. G. Littlejohn: *Non-generic spectral statistics in the quantized stadium billiard*, J. Phys. A **26** (1993) 6217–6230.

41. A. Bäcker, F. Steiner and P. Stifter: *Spectral statistics in the quantized cardioid billiard*, Phys. Rev. E **52** (1995) 2463–2472.

42. A. Bäcker and F. Steiner: *Quantum chaos and quantum ergodicity*, in *Ergodic Theory, Analysis and Efficient Simulation of Dynamical Systems*, B. Fiedler (ed.), 717- 752, Springer-Verlag Berlin/Heidelberg (2001).

43. J. R. Kuttler and V. G. Sigilito: *Eigenvalues of the Laplacian in two dimensions*, SIAM Review **26** (1984) 163–193.

44. E. J. Heller: *Wavepacket dynamics and quantum chaology*, in: *Proceedings of the 1989 Les Houches School on Chaos and Quantum Physics* (Eds. M.-J. Giannoni, A. Voros and J. Zinn Justin), North–Holland, Amsterdam, (1991).

45. B. Li and M. Robnik: *Statistical properties of high-lying chaotic eigenstates*, J. Phys. A **27** (1994) 5509–5523.

46. E. Doron and U. Smilansky: *Chaotic Spectroscopy*, Chaos **2** (1992) 117–124.

47. B. Dietz and U. Smilansky: *A scattering approach to the quantization of billiards – The inside-outside duality*, Chaos **3** (1993) 581–590.

48. H. Schanz and U. Smilansiky: *Quantization of Sinai's billiard – a scattering approach*, Chaos, Solitons and Fractals **5** (1995) 1289–1309.

49. E. Vergini and M. Saraceno: *Calculation of highly excited states of billiards*, Phys. Rev. E **52** (1995) 2204–2207.
50. A. J. Burton and G. F. Miller: *The application of integral equation methods to the numerical solution of some exterior boundary-value problems*, Proc. R. Soc. London Ser. A **323** (1971) 201–210.
51. R. E. Kleinman and G. F. Roach: *Boundary integral equations for the three dimensional Helmholtz equation*, SIAM Rev. **16** (1974) 214–236.
52. R. J. Riddel Jr.: *Boundary-distribution solution of the Helmholtz equation for a region with corners*, J. Comp. Phys. **31** (1979) 21–41.
53. R. J. Riddel Jr.: *Numerical solution of the Helmholtz equation for two-dimensional polygonal regions*, J. Comp. Phys. **31** (1979) 42–59.
54. P. A. Martin: *Acoustic scattering and radiation problems and the null-field method*, Wave Motion (1982) 391–408.
55. M. V. Berry and M. Wilkinson: *Diabolical points in the spectra of triangles*, Proc. R. Soc. London Ser. A **392** (1984) 15–43.
56. M. Sieber and F. Steiner: *Quantum chaos in the hyperbola billiard*, Phys. Lett. A **148** (1990) 415–419.
57. D. Biswas and S. Jain: *Quantum description of a pseudointegrable system: the π/3-rhombus billiard*, Phys. Rev. A **42** (1990) 3170–3185.
58. P. A. Boasmann: *Semiclassical Accuracy for Billiards*, Ph.D. thesis, H. H. Wills Physics Laboratory, Bristol, (1992).
59. P. A. Boasmann: *Semiclassical accuracy for billiards*, Nonlinearity **7** (1994) 485–537.
60. R. Aurich and F. Steiner: *Statistical properties of highly excited quantum eigenstates of a strongly chaotic system*, Physica D **64** (1993) 185–214.
61. C. Pisani: *Exploring periodic orbit expansions and renormalisation with the quantum triangular billiard*, Ann. Physics **251** (1996) 208–265.
62. I. Kosztin and K. Schulten: *Boundary integral method for stationary states of two-dimensional quantum systems*, Int. J. Mod. Phys. C **8** (1997) 293–325.
63. B. Li, M. Robnik and B. Hu: *Relevance of chaos in numerical solutions of quantum billiards*, Phys. Rev. E **57** (1998) 4095–4105.
64. M. Sieber: *Billiard systems in three dimensions: the boundary integral equation and the trace formula*, Nonlinearity **11** (1998) 6 1607–1623.
65. K. Hornberger and U. Smilansky: *The boundary integral method for magnetic billiards*, J. Phys. A **33** (1999) 2829–2855.
66. R. Aurich and J. Marklof: *Trace formulae for three-dimensional hyperbolic lattices and application to a strongly chaotic tetrahedral billiard*, Physica D **92** (1996) 101–129.
67. H. Primack and U. Smilansky: *Quantization of the 3-dimensional Sinai billiard*, Phys. Rev. Lett. **74** (1995) 4831–4834.
68. G. Steil: *Eigenvalues of the Laplacian for Bianchi groups*, in: *Emerging applications of number theory (Minneapolis, MN, 1996)*, 617–641, Springer, New York, (1999).
69. T. Prosen: *Quantization of generic chaotic 3D billiard with smooth boundary I: energy level statistic*, Phys. Lett. A **233** (1997) 323–331.
70. T. Prosen: *Quantization of generic chaotic 3D billiard with smooth boundary II: structure of high-lying eigenstates*, Phys. Lett. A **233** (1997) 332–342.
71. R. D. Ciskowski and C. Brebbia, eds.: *Boundary Element Methods in Acoustics*. Computational Mechanics Publications and Elsevier Applied Science, (1991).

72. E. B. Bogomolny: *Semiclassical quantization of multidimensional systems*, Nonlinearity **5** (1992) 805–866.

73. T. Harayama and A. Shudo: *Zeta function derived from the boundary element method*, Phys. Lett. A **165** (1992) 417–426.

74. B. Burmeister: *Korrekturen zur Gutzwillerschen Spurformel für Quantenbillards*, Diploma Thesis , II. Institut für Theoretische Physik, Universität Hamburg (1995).

75. M. Sieber, N. Pavloff and C. Schmit: *Uniform approximation for diffractive contributions to the trace formula in billiard systems*, Phys. Rev. E **55** (1997) 2279–2299.

76. B. Burmeister and F. Steiner: *Exact trace formula for quantum billiards*, unpublished (1995).

77. S. Tasaki, T. Harayama and A. Shudo: *Interior Dirichlet eigenvalue problem, exterior Neumann scattering problem, and boundary element method for quantum billiards*, Phys. Rev. E **56** (1997) R13–R16.

78. J.-P. Eckmann and C.-A. Pillet: *Zeta functions with Dirichlet and Neumann boundary conditions for exterior domains*, Helv. Phys. Acta **70** (1997) 44–65.

79. R. Aurich: private communication.

80. M. Robnik: *Quantising a generic family of billiards with analytic boundaries*, J. Phys. A **17** (1984) 1049–1074.

81. T. Prosen and M. Robnik: *Energy level statistics and localization in sparsed banded random matrix ensembles*, J. Phys. A **26** (1993) 1105–1114.

82. M. Abramowitz and I. A. Stegun (eds.): *Pocketbook of Mathematical Functions*, Verlag Harri Deutsch, Thun – Frankfurt/Main, abridged edn., (1984).

83. A. Sommerfeld: *Vorlesungen über Theoretische Physik, Band VI: Partielle Differentialgleichungen der Physik*, Harri Deutsch, Thun, (1984).

84. T. Hesse: *Semiklassische Untersuchung zwei– und dreidimensionaler Billardsysteme*, Ph.D. thesis, Abteilung Theoretische Physik, Universität Ulm, (1997).

85. K. Życzkowski: *Classical and quantum billiards, integrable, nonintegrable, and pseudo-integrable*, Acta Physica Polonica B **23** (1992) 245–270.

86. A. Bäcker and R. Schubert: *Chaotic eigenfunctions in momentum space*, J. Phys. A **32** (1999) 4795–4815.

87. J. M. Tualle and A. Voros: *Normal modes of billiards portrayed in the stellar (or nodal) representation*, Chaos, Solitons and Fractals **5** (1995) 1085–1102.

88. F. P. Simonotti, E. Vergini and M. Saraceno: *Quantitative study of scars in the boundary section of the stadium billiard*, Phys. Rev. E **56** (1997) 3859–3867.

89. A. Bäcker and R. Schubert: *Autocorrelation function of eigenstates in chaotic and mixed systems*, J. Phys. A **35** (2002) 539–564.

90. E. J. Heller: *Bound-state eigenfunctions of classically chaotic Hamiltonian systems: Scars of periodic orbits*, Phys. Rev. Lett. **53** (1984) 1515–1518.

91. Python, http://www.python.org/.

92. Numerical Python (NumPy), http://sourceforge.net/projects/numpy/.

93. SciPy, http://www.scipy.org/.

94. Gnuplot, http://www.gnuplot.info/.

From Normal to Anomalous Deterministic Diffusion

Roberto Artuso

Centre for Nonlinear and Complex Systems and Dipartimento di Scienze Chimiche Fisiche, Chimiche e Matematiche and I.N.F.M. Sezione di Como, Via Valleggio, 11 I-22100 Como, Italy, *roberto.artuso@uninsubria.it*

Summary. These lecture notes illustrate some features of deterministic transport in chaotic systems. The subject has witnessed an impressive amount of work in the last thirty years, and our review is not meant to be exhaustive, but rather focus on some unifying techniques by which the problem can be tackled, pointing out difficulties and open problems.

We start by dealing with the case of *hyperbolic* systems where typically normal diffusion is observed (even though actual calculation of transport coefficients may be exceedingly difficult), while the second part of the notes deals with *weakly chaotic* systems, where long trappings near regular phase-space regions may induce anomalies in diffusive properties. Examples of analytic calculations are given in the framework of *cycle expansions*, a general technique for getting chaotic averages.

1 Normal Diffusion

1.1 One-Dimensional Maps on the Real Line

We start by considering completely unstable dynamical systems, and our prototype example will be one dimensional lifts of piecewise linear maps on the interval. First we define the map $\hat{f}$ on the unit interval, in the following way

$$\hat{f}(\hat{x}) = \begin{cases} \Lambda\hat{x} & \hat{x} \in [0, 1/2] \\ \Lambda\hat{x} + 1 - \Lambda & \hat{x} \in (1/2, 1] \end{cases} \tag{1}$$

and then lift it to the real line, through the translation property

$$\hat{f}(\hat{x} + n) = \hat{f}(\hat{x}) + n \qquad n \in \mathbf{Z} \tag{2}$$

The former definition guarantees that also the symmetry property

$$\hat{f}(\hat{x}) = -\hat{f}(-\hat{x}) \tag{3}$$

is satisfied, so that the dynamics will not present any drift.

The translational symmetry (2) relates the unbounded dynamics on the real line to the dynamics restricted to a *fundamental cell* - in the present example the unit interval curled up into a circle.

Associated to $\hat{f}(\hat{x})$ we thus also consider the circle map

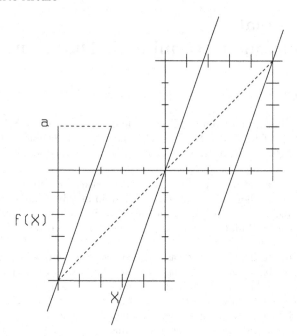

Fig. 1. The full sawtooth map (1), $a = \Lambda/2$.

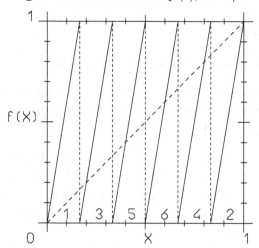

Fig. 2. The corresponding circle map (4). ($\Lambda = 4$)

$$f(x) = \hat{f}(\hat{x}) - \left[\hat{f}(\hat{x})\right] , \qquad x = \hat{x} - [\hat{x}] \in [0,1] \tag{4}$$

(see Fig. 2), where $[\cdots]$ stands for the integer part. In our treatment a key role will be played by periodic orbits of the dynamical system: we see here that periodic orbits of the torus map are of two types: a periodic orbit $p =$

$\{x_1, x_2, \ldots x_n\}$, $f^n x_j = x_j$, is called *standing* if it is also a periodic orbit of the dynamics on the line, $\hat{f}^n(\hat{x}_j) = \hat{x}_j$, while it is called *running* if it correspond to a translation in the dynamics on the line (in the theory of area–preserving maps such orbits are usually called *accelerator modes*), $\hat{f}^n(x_j) = x_j + \hat{n}_p$. We shall refer to $\hat{n}_p \in \mathbf{Z}$ as the *jumping number* of the p cycle.

Though maps (1) have a particularly simple form, transport properties are not trivial: Klages and Dorfman [1] have shown how the diffusion coefficient D is a fractal function of the slope Λ.

1.2 Transfer Operator for Diffusion

We now briefly recall how the diffusion coefficient D may be obtained by taking into account periodic orbits of the (torus) map [2–4] (see also [5,6]). Our analysis is not confined to simple maps of the form (1): the only requirements will be that symmetry properties (2,3) hold, as well as hyperbolicity of the map. We recall that the diffusion coefficient is defined as

$$\langle (\hat{x}_n - \hat{x}_0)^2 \rangle_{\mathcal{I}} \sim 2D \cdot n \tag{5}$$

where $\langle \ldots \rangle_{\mathcal{I}}$ means average with respect to a distribution of initial conditions (typically uniformly chosen over a cell). A useful object (which contains information about every moment of the distribution) is the generating function:

$$\mathcal{G}_n(\beta) = \langle e^{\beta(\hat{x}_n - \hat{x})} \rangle_{\mathcal{I}} \tag{6}$$

We want to introduce a suitable transfer operator through which asymptotic behavior of the generating function may be expressed: this task has to face a problem, which is apparent if we just look at the first iterate

$$\mathcal{G}_1(\beta) = \int_{\mathcal{I}} dx \int_{\mathbf{R}} dy \, e^{\beta(\hat{f}(x) - x)} \delta(y - \hat{f}(x)) \tag{7}$$

so that the "indices" of the kernel are mismatched, coming from two different sets ($\mathcal{I}$ and $\mathbf{R}$). Now take $\mathcal{I} = [0, 1]$: we may observe that $\forall y \in \mathbf{R}$, $y = \hat{f}(x)$ there exists one and only one $z \in \mathcal{I}$ such that $z = f(x)$ and $y = z + n_x$, $n_x \in \mathbf{Z}$, so that (7) may be rewritten as

$$\mathcal{G}_1(\beta) = \int_{\mathcal{I}} dx \int_{\mathcal{I}} dz \, e^{\beta(f(x) + n_x - x)} \delta(z - f(x)) \tag{8}$$

In this way transport properties are encoded by the *torus* map f, together with the set of "jumping numbers" n_x. Now we may introduce a *generalized transfer operator* (enjoying a semigroup property) $\mathcal{L}_\beta$ as

$$(\mathcal{L}_\beta h)(x) = \int_{\mathcal{I}} dz \, e^{\beta(f(z) + n_z - z)} \delta(x - f(z)) \tag{9}$$

whose (singular) integral kernel is

$$\mathcal{L}_\beta(x, y) = e^{\beta(f(y)+n_y-y)} \delta(x - f(y)) \tag{10}$$

If we now ignore the fact that the integral kernel is singular (so that $\mathcal{L}_\beta$ is not compact like in ordinary Fredholm theory) and label eigenvalues of (9) in decreasing order (with respect to their absolute value) $\lambda_0(\beta), \lambda_1(\beta), \ldots$ we may formally write, for large n

$$\mathcal{G}_n(\beta) = \int_\mathcal{I} dx \int_\mathcal{I} dy \, \mathcal{L}_\beta^n(x, y) \sim \sum_{j=0} \lambda_j^n \tag{11}$$

In particular if the leading eigenvalue is isolated (*i.e.* some generalized Perron theorem holds) the generating function is asymptotically dominated by powers of $\lambda_0(\beta)$; on the other hand, by power expansion in β we have that

$$\mathcal{G}_n(\beta) = 1 + \beta \langle (\hat{x}_n - \hat{x}_0) \rangle_\mathcal{I} + \frac{\beta^2}{2!} \langle (\hat{x}_n - \hat{x}_0)^2 \rangle_\mathcal{I} \tag{12}$$

and we may thus relate the diffusion constant to the leading eigenvalue of the generalized transfer operator in the following way

$$D = \frac{1}{2} \frac{d^2 \lambda_0(\beta)}{d\beta^2}\bigg|_{\beta=0} \tag{13}$$

1.3 Cycle Expansion for D

In view of (13) the diffusion coefficient may be computed once we know the leading eigenvalue of the generalized transfer operator (9), $\lambda_0(\beta)$, whose inverse $z(\beta)$ is the smallest root of the equation

$$\det(1 - z\mathcal{L}_\beta) = 0 \tag{14}$$

We now see how to relate $z(\beta)$ to properties of periodic orbits of the torus map.

Step One: Going to Traces.

We use the standard formula relating the logarithm of determinant to traces

$$\det(1 - z\mathcal{L}_\beta) = \exp \ln \det(1 - z\mathcal{L}_\beta) = \exp - \sum_{k=1}^\infty \frac{z^k}{k} \mathrm{tr} \mathcal{L}_\beta^k \tag{15}$$

where

$$\mathrm{tr} \mathcal{L}_\beta = \int_\mathcal{I} dz \, \mathcal{L}_\beta(z, z) \tag{16}$$

Step Two: Traces and Periodic Orbits.

Now we use properties of the Dirac distribution to see how traces are expressed in terms of periodic points sums

$$\text{tr}\mathcal{L}_\beta^k = \sum_{z|f^k(z)=z} \frac{e^{\beta n_{(k),z}}}{\left|1 - \Lambda_{(k),z}\right|} \tag{17}$$

where

$$n_{(k),z} : \hat{f}^k(z) = z + n_{(k),z} \tag{18}$$

and the contribution of different periodic points are weighted by their stability

$$\Lambda_{(k),z} = \frac{d^k f}{dz^k}(z) = \prod_{m=0}^{k-1} f'(f^m(z)) \tag{19}$$

Step Three: A Few Key Observations.

- Both the jumping factors $n_{(k),z}$ and the stability $\Lambda_{(k),z}$ are the same for all periodic points belonging to the same periodic orbit.
- For each order k (17) picks up contributions from both orbits of "prime period" k, and by orbits of smaller periods s such that s divides k (in particular fixed points contribute to all orders). Now suppose that z is a point of a periodic orbit of prime period s and $k = s \cdot m$, then we have $n_{(k),z} = m \cdot n_{(s),z}$ and $\Lambda_{(k),z} = \Lambda_{(s),z}^m$ thus the only independent quantities entering the former expressions are stabilities and jumping numbers of prime cycles.

Step Four: Resumming Out Repetitions.

We now take explicitly into account the hypothesis of full hyperbolicity $|\Lambda_{(m),z}| > 1$, so that the denominator in (17) may always be expanded as a geometric series: we get

$$\sum_{k=1}^\infty \frac{z^k}{k} \sum_{z|f^k(z)=z} \frac{e^{\beta n_{(k),z}}}{\left|1 - \Lambda_{(k),z}\right|} = \sum_{\{p\}} \sum_{j=0}^\infty \sum_{r=1}^\infty \frac{n_p z^{n_p \cdot r}}{n_p \cdot r} \frac{e^{\beta r \sigma_p}}{|\Lambda_p|^r \Lambda_p^{j \cdot r}} \tag{20}$$

where $\sigma_p = n_{(n_p),z_p}$ (z_p being any point of the periodic orbit labelled by p) and, analogously, $\Lambda_p = \Lambda_{(n_p),z_p}$. In (20) $\{p\}$ indicates a sum over all "prime periodic orbits" (of prime period n_p), j is the geometric series index coming from expanding the denominators, while r counts repetitions of prime cycles (as in the original sum each p cycle appears at every $r \cdot n_p$ order). Now we may sum up the r (logarithmic) series, thus getting

$$\det(1 - z\mathcal{L}_\beta) = \exp \sum_{\{p\}} \sum_{j=0}^\infty \ln\left(1 - z^{n_p} \frac{e^{\beta \cdot \sigma_p}}{|\Lambda_p|\Lambda_p^j}\right) = \prod_{j=0}^\infty \zeta_{\beta,(j)}^{-1}(z) \tag{21}$$

where *dynamical zeta functions* are thus defined as

$$\zeta_{\beta,(j)}^{-1}(z) = \prod_{\{p\}} \left(1 - z^{n_p} \frac{e^{\beta \cdot \sigma_p}}{|\Lambda_p| \Lambda_p^j} \right) \tag{22}$$

Generically the smallest zero $z(\beta)$ (whose inverse provides the leading eigenvalue of the generalized transfer operator) comes from the lowest order zeta function [7],

$$\zeta_{\beta,(0)}^{-1}(z) = \prod_{\{p\}} \left(1 - z^{n_p} \frac{e^{\beta \cdot \sigma_p}}{|\Lambda_p|} \right) \tag{23}$$

Cycle expansions [7–9] consist in turning (23) into a power series

$$\zeta_{\beta,(0)}^{-1}(z) = 1 - \sum_{m=1}^{\infty} \gamma_m(\beta) z^m \tag{24}$$

Finite l-order estimates (that require information from periodic orbits whose prime period does not exceed l) come from polynomial truncation of (24): this leads to a genuine perturbation scheme if we are able to control how finite order estimates converge to the asymptotic limit. This is strictly related to establishing *analytic* properties of the dynamical zeta function (typically that ζ is meromorphic in some disc): this subject is outside the scope of our review: we refer to [9] (for euristic arguments and explicit examples) and [10] (for a rigorous approach, and a guide to the relevant mathematical references).

1.4 A Simple Example

We return to the map (1), and take $\Lambda = 6$. The corresponding torus map is shown in Fig. 2: it consists of six branches, each mapping its support $(\mathcal{I}_k, k = 1, \ldots, 6)$ *onto* the whole interval. This leads to a very simple symbolic dynamics (a simple introduction to symbolic dynamics is contained in [11], the importance of its control in the framework of cycle expansion is discussed in [9]: advanced topics are discussed in [12]). More precisely every infinite combination of six letters (labelling the sets $\mathcal{I}_k$) corresponds to a different orbit of the dynamical system, and in particular periodic sequences individuate periodic points. The diffusing properties of the corresponding map on the real line are encoded by the jumping factors, that (see (1)) are given as follows

$$\sigma_1 = \sigma_2 = 0 \qquad \sigma_3 = -\sigma_4 = +1 \qquad \sigma_5 = -\sigma_6 = +2 \tag{25}$$

moreover to each symbol a stability $\Lambda_k = \Lambda = 6$ is associated. As the slope is uniform, and jumping factors are uniquely determined by the symbol, for each prime periodic orbit we have

$$\sigma_p = \sigma_{\epsilon_1, \epsilon_2 \ldots \epsilon_{n_p}} = \sum_{k=1}^{n_p} \sigma_{\epsilon_k} \qquad \Lambda_p = \Lambda^{n_p} \tag{26}$$

and the zeta function (23) is a first order polynomials [7,9], being determined by fixed points only

$$\zeta_{\beta,(0)}^{-1}(z) = 1 - 2\frac{z}{\Lambda} - 2z\frac{\cosh(\beta)}{\Lambda} - 2z\frac{\cosh(2\beta)}{\Lambda} \tag{27}$$

The smallest (and only) zero is thus easily found as

$$z(\beta) = \lambda_0(\beta)^{-1} = \frac{\Lambda}{2 \cdot (1 + \cosh(\beta) + \cosh(2\beta))} \tag{28}$$

from which we get by (13) the exact expression of the diffusion coefficient

$$D = \frac{1}{2}\frac{d^2\lambda_0(\beta)}{d\beta^2}\bigg|_{\beta=0} = \frac{5}{6} \tag{29}$$

Actually a careful analysis of the symbolic dynamics of this class of maps shows that there is an infinite number of Λ values for which an exact expression for D can be analytically computed (namely every time the symbolic dynamics is generated by a finite Markov graph [9]). However this property is strictly related to number theoretical properties of Λ: solvable cases are embedded in the range of Λ values is the same way rationals are embedded in the unit interval: it is precisely this extremely subtle dependence of the symbolic dynamics on Λ that makes the behavior of $D(\Lambda)$ so complex [1].

1.5 A Nontrivial Example

In this section (where we follow mainly [13]) we will deal with the prototype example of chaotic Hamiltonian maps, hyperbolic toral automorphisms (Anosov maps) [14]. Diffusive properties will arise in considering such maps acting on the cylinder or over $\mathbf{R}^2$, while the dynamics restricted to the fundamental domain involves maps on $\mathbf{T}^2$ (two–dimensional torus). An Anosov map thus corresponds to the action of a matrix in $SL_2(\mathbf{Z})$ with unit determinant and absolute value of the trace bigger than 2.

Maps of this kind are as examples of genuine Hamiltonian chaotic evolution. They admit simple finite Markov partitions [11, 14], which lead to simple analytic expressions for ordinary ($\beta = 0$) zeta functions [15]. Within the framework of Hamiltonian dynamical systems their role is analogous to piecewise linear Markov maps: their symbolic dynamics can be encoded in a simple way, and their linearity leads to uniformity of cycle stabilities.

We will consider the "two-coordinates" representation for them

$$\begin{bmatrix} x' \\ y' \end{bmatrix} = M \begin{bmatrix} x \\ y \end{bmatrix}$$

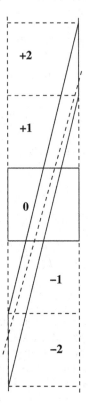

Fig. 3. Symbolic dynamics for the cat map (Percival Vivaldi coding).

with

$$M = \begin{bmatrix} 0 & 1 \\ -1 & K+2 \end{bmatrix}$$

(see Fig. 3), on $[-1/2, 1/2)^2$, extending the map on the cylinder $[-1/2, 1/2) \times \mathbf{R}$ through symmetry property (2).

Though Markov partitions encode the symbolic dynamics in the simplest possible way, they are not well suited to deal with diffusion, as the jumping factor is not related in a simple way to the orbit symbol sequence. To this end the linear code introduced by Percival and Vivaldi [16, 17] is quite natural: before describing it let us fix the notations: χ will denote the trace of the map ($\chi = K + 2$): the leading eigenvalue will de denoted by $\lambda = (\chi + \sqrt{\ell})/2$, where $\ell = \chi^2 - 4$. In principle the code (and the problem of diffusion) can be also considered for real values of K (thus loosing continuity of the torus map when K in not an integer): we will remark in what follows as results which are exact for $K \in \mathbf{N}$ are only approximate for generic K.

The cardinality of the alphabet is determined by the parameter K: the letters are integer numbers (giving the integer jumps of the trajectory (see

Fig. 3), whose absolute values does not exceed $Int(1 + \chi/2)$. The code is linear, as, given a bi-infinite sequence $\{x_i\}_{i\in\mathbf{Z}}$

$$b_t \stackrel{\text{def}}{=} \left[(K+2)x_t - x_{t-1} + \frac{1}{2}\right] \quad , \tag{30}$$

[...] denoting the integer part, while the inversion formula (once a bi-infinite symbolic string $\{b_i\}_{i\in\mathbf{Z}}$ is given), reads

$$x_t = \frac{1}{\sqrt{\ell}} \sum_{s\in\mathbf{Z}} \lambda^{-|t-s|} b_s \quad , \tag{31}$$

As the x coordinate lives in the interval $[-1/2, 1/2)$, (31) induces a condition on allowed symbol sequences: $\{b_i\}_{i\in\mathbf{Z}}$ corresponds to an admissible orbit if

$$\frac{1}{2} \leq \frac{1}{\sqrt{\ell}} \sum_{s\in\mathbf{Z}} \lambda^{-|t-s|} b_s < \frac{1}{2} \quad . \tag{32}$$

By (30,31) it is easy to observe that periodic orbits and allowed periodic symbol sequences are in one-to-one correspondence. From (32) we get the condition that a $\{b_i\}_{i=1,\ldots,T}$ sequence corresponds to a T–periodic orbit of the map

$$|A_n b_t + A_{n-1}(b_{t+1} + b_{t-1}) + \cdots + A_0(b_{t+n} + b_{t-n})| < \frac{B_n}{2} \quad \forall t = 1,\ldots,T$$

when $T = 2n + 1$, and

$$|C_n b_t + C_{n-1}(b_{t+1} + b_{t-1}) + \cdots + C_0(b_{t+n})| < \frac{D_n}{2} \quad \forall t = 1,\ldots,T \tag{34}$$

when $T = 2n$ where

$$B_k = \lambda^k(\lambda - 1) + \lambda^{-k}(\lambda^{-1} - 1) \qquad A_k = \frac{\lambda^{k+1} + \lambda^{-k}}{\lambda + 1}$$
$$D_k = (\lambda^k - \lambda^{-k})(\lambda - \lambda^{-1}) \qquad C_k = \lambda^k + \lambda^{-k} \tag{35}$$

We notice that these numbers satisfy the following recursion relations

$$u_{k+2} = \chi u_{k+1} - u_k$$

where $u_k = A_k, B_k, C_k, D_k$ (in particular for $\chi = 3$, $A_k = F_{2k+1}$, $B_k = L_{2k+1}$, $C_k = L_{2k}$ and $D_k = 5F_{2k}$, where F_n and L_n are the Fibonacci and Lucas numbers).

The pruning rules (34) admit a simple geometric interpretation: a lattice point $\mathbf{b} \in \mathbf{Z}^{\mathbf{T}}$ individuates a T–periodic point of the map if $\mathbf{b} \in \mathcal{P}_{\mathbf{T}}$ where

$$\mathcal{P}_T \stackrel{\text{def}}{=} \left\{ \mathbf{x} \in \mathbf{R}^{\mathbf{T}} : \begin{array}{c} |a_1 x_1 + \cdots + a_T x_T| < e_T \\ \vdots \\ |a_2 x_1 + \cdots + a_1 x_T| < e_T \end{array} \right\} \tag{36}$$

and

$$a_1 \ldots a_T = A_0 A_1 \ldots A_{n-1} A_n A_{n-1} \ldots A_0 \qquad e_T = B_n/2$$
$$a_1 \ldots a_T = C_1 \ldots C_{n-1} C_n C_{n-1} \ldots C_1 C_0 \qquad e_T = D_n/2 \tag{37}$$

for $T = 2n + 1$ or $T = 2n$, respectively, Thus $\mathcal{P}_T$ is a measure polytope [18], obtained by deforming a T-cube.

We will denote by $\mathcal{N}_{n,s}$ the number of periodic points of period n with jumping number s. By taking into account (17)), and observing that

$$\mathrm{tr}\mathcal{L}_\beta^n \sim \lambda_0(\beta)^n \tag{38}$$

we can easily see that for cat maps a way to compute D is provided by

$$D = \lim_{n \to \infty} D_n \qquad D_n = \frac{1}{n\mathcal{N}_n} \sum_{k=1}^{p(n)} k^2 \mathcal{N}_{n,k} \tag{39}$$

where $\mathcal{N}_n$ is the number of periodic points of period n, $p(n)$ is the highest jumping number of n–periodic orbits and we employed

$$\left| \det \left(1 - \mathbf{J}_{\mathbf{x}}^{(n)} \right) \right| = (\lambda^n - 1)(1 - \lambda^{-n}) = \mathcal{N}_n$$

which is valid for cat maps (the jacobian matrix appears as two-dimensional generalization of stability).

Sums can be converted into integrals by using Poisson summation formula: we define

$$f_T(n) = \begin{cases} (n_1 + \cdots + n_T)^2 & n \in \mathcal{P}_T \cap \mathbf{Z}^T \\ 0 & otherwise \end{cases}$$

and

$$\tilde{f}_T(\xi) = \int_{\mathbf{R}^T} dx\, e^{i(x,\xi)} f_T(x)$$

From Poisson summation formula we have that

$$D_T = \frac{1}{T\mathcal{N}_T} \sum_{n \in \mathbf{Z}^T} \tilde{f}_T(2\pi n) \tag{40}$$

The quasilinear estimate for D_T (see for example [19]) amounts to considering the $n = 0$ contribution to (40):

$$D_T^{(q.l.)} = \int_{\mathcal{P}_T} dx\, (x_1 + x_2 + \cdots + x_T)^2 \tag{41}$$

The evaluation of (41) requires introducing a coordinate transformation in symbolic space in which $\mathcal{P}_T$ is transformed in a T–cube. This is equivalent to finding the inverse of the matrix A:

$$A \overset{\text{def}}{=} \begin{pmatrix} a_1 & a_2 & \cdots & a_{T-1} & a_T \\ a_T & a_1 & \cdots & a_{T-2} & a_{T-1} \\ \vdots & \vdots & \ddots & \vdots & \vdots \\ a_3 & a_4 & \cdots & a_1 & a_2 \\ a_2 & a_3 & \cdots & a_T & a_1 \end{pmatrix}. \tag{42}$$

First of all let us observe that A is a circulant matrix, so that its determinant is the product of T factors, each of the form $f(\epsilon_j) = a_1 + \epsilon_j a_2 + \cdots + a_T \epsilon_j^{T-1}$, where ϵ_j is a T-th root of unity. By using (35) it is possible to see that

$$f(\epsilon_j) = \begin{cases} \dfrac{\epsilon_j^{n+1} B_n}{(\lambda \epsilon_j - 1)(1 - \lambda^{-1} \epsilon_j)} & T = 2n+1 \\ \dfrac{\epsilon_j^{n} D_n}{(\lambda \epsilon_j - 1)(1 - \lambda^{-1} \epsilon_j)} & T = 2n \end{cases}$$

so that

$$|\det A| = \frac{(2e_T)^T}{\lambda^T + \lambda^{-T} - 2} \tag{43}$$

A can be diagonalized [13] by considering the auxiliary matrix U

$$U \overset{\text{def}}{=} \begin{pmatrix} 1 & 1 & \cdots & 1 & 1 \\ \epsilon_0 & \epsilon_1 & \cdots & \epsilon_{T-2} & \epsilon_{T-1} \\ \vdots & \vdots & \ddots & \vdots & \vdots \\ \epsilon_0^{T-2} & \epsilon_1^{T-2} & \cdots & \epsilon_{T-2}^{T-2} & \epsilon_{T-1}^{T-2} \\ \epsilon_0^{T-1} & \epsilon_1^{T-1} & \cdots & \epsilon_{T-2}^{T-1} & \epsilon_{T-1}^{T-1} \end{pmatrix}.$$

In fact $U^{-1} A U$ is a diagonal matrix (the diagonal elements coinciding with $f(\epsilon_j)$).

We can finally express A^{-1} via

$$\tilde{C} A^{-1} = \frac{1}{B_n^T} \begin{pmatrix} \chi & -1 & \cdots & 0 & -1 \\ -1 & \chi & \cdots & 0 & 0 \\ \vdots & \vdots & \ddots & \vdots & \vdots \\ 0 & 0 & \cdots & \chi & -1 \\ -1 & 0 & \cdots & -1 & \chi \end{pmatrix} \tag{44}$$

where

$$\tilde{C} = \begin{pmatrix} \mathbf{0} & \mathbf{1}_{n+1} \\ \mathbf{1}_n & \mathbf{0} \end{pmatrix}.$$

if $T = 2n+1$ and

$$\tilde{K} A^{-1} = \frac{1}{D_n^T} \begin{pmatrix} \chi & -1 & \cdots & 0 & -1 \\ -1 & \chi & \cdots & 0 & 0 \\ \vdots & \vdots & \ddots & \vdots & \vdots \\ 0 & 0 & \cdots & \chi & -1 \\ -1 & 0 & \cdots & -1 & \chi \end{pmatrix} \tag{46}$$

where

$$\tilde{K} = \begin{pmatrix} 0 & 1_n \\ 1_n & 0 \end{pmatrix} \; .$$

if $T = 2n$. As a first check of quasilinear estimates let's compute the volume of $\mathcal{P}_T$:

$$Vol(\mathcal{P}_T) = \int_{\mathcal{P}_T} dx_1 \, dx_2 \ldots dx_T = \frac{1}{|\det A|} \int_{-e_T}^{e_T} \ldots \int_{e_T}^{e_T} d\xi_1 \ldots d\xi_T$$

$$= \lambda^T + \lambda^{-T} - 2$$

which is exactly the number of T–periodic points of the map when K is an integer.

In an analogous way we may compute the quasilinear estimate for $\mathcal{N}_{T,k}$

$$\mathcal{N}_{T,k}^{(q.l.)} = \int_{\mathcal{P}_T} dx_1 \ldots dx_T \, \delta(x_1 + \ldots x_T - k)$$

$$= \frac{\lambda^T + \lambda^{-T} - 2}{(2e_T)^T} \int_{-\infty}^{\infty} d\alpha \, e^{-2\pi i \alpha k} \int_{-e_T}^{e_T} \ldots \int_{-e_T}^{e_T} d\xi_1 \ldots d\xi_T \, e^{\frac{2\pi i \alpha \chi}{2e_T}(\xi_1 + \cdots + \xi_T)}$$

$$= \frac{2}{\pi \chi}(\lambda^T + \lambda^{-T} - 2) \int_0^{\infty} dy \, \cos\left(\frac{2qy}{\chi}\right) \left(\frac{\sin y}{y}\right)^T \tag{48}$$

where we have used $x_1 + \cdots + x_T = (\chi/(2e_T))(\xi_1 + \cdots + \xi_T)$ ((44,46)).

We are now ready to evaluate the quasilinear estimate for the diffusion coefficient

$$D_T^{(q.l.)} = \frac{1}{\pi \chi T} \int_{-T\chi/2}^{T\chi/2} dz \, z^2 \int_0^{\infty} dy \, \cos\left(\frac{2zy}{\chi}\right) \left(\frac{\sin y}{y}\right)^T \tag{49}$$

(where the bounds on the jumping number again come easily from (44,46)). By dropping terms vanishing as $T \mapsto \infty$, and using

$$\int_0^{\infty} dx \left(\frac{\sin x}{x}\right)^n \frac{\sin(mx)}{x} = \frac{\pi}{2} \quad m \geq n$$

we can evaluate

$$D^{(q.l.)} = \frac{\chi^2}{24} \tag{50}$$

which is the correct result [20] (and again for cat maps (50) is not the quasilinear estimate but the exact value of the diffusion coefficient).

2 Intermittency and Anomalous Diffusion

In our treatment of normal diffusion the key assumption was a hyperbolicity property: all the periodic orbits have to be unstable. For a wide class of relevant dynamical systems ("generic" Hamiltonian systems, Lorentz gas with infinite horizon) this hypothesis fails and the sticking of trajectories to stable islands influences the global properties of the dynamics, *e.g.* by making the diffusion anomalous or inducing power–law decay of correlations. We will try to provide evidences that the way marginal stability must be accounted for by (diffusion) cycle expansions is by gauging the role of longer and longer unstable orbits accumulating to marginally stable cycles.

We begin by discussing some simple one-dimensional maps almost everywhere hyperbolic (non hyperbolicity appearing only in a marginally stable fixed point): this will allow us to reformulate the cycling approach to diffusion in order to include anomalous transport properties. In this context we will introduce an approximation leading to probabilistic dynamical zeta functions, which in particular allows a rather accurate analysis of dynamical properties of the Lorentz gas with infinite horizon.

2.1 An Intermittent Map

Before dealing with transport properties we briefly introduce a typical weakly chaotic map, pointing out the main difficulties that arise when describing its dynamics through zeta functions. The map has the following form [21]

$$x_{n+1} = f(x_n) = x_n + cx_n^z \ (mod\,1) \tag{1}$$

($z > 1$, $c > 0$), which is manifestly non–hyperbolic, since $x = 0$ is a marginal fixed point (indifferent equilibrium point). We take $c = 1$, so that the map consists of two full branches, with support on $[0, p)$ (I_0) and $[p, 1]$ (I_1), where $p + p^z = 1$ (see Fig. **??**). The inverse of $f|_{I_i}$ will be denoted by ϕ_i.

The presence of two full branches (each mapping its support onto the whole unit interval) apparently suggests the use of an unrestricted binary alphabet to encode the dynamics: but the $\bar{0}$ fixed point is marginal and cannot be included directly in cycle expansions, its role being probed by infinity of cycles accumulating to it. The pruning of $\bar{0}$ is achieved by considering the following (countable) new alphabet: $\{1, 0^k1, \ k = 1, 2\ldots\}$. The unit interval is accordingly partitioned into a sequence of subsets I_1, I_{0^k1}, where $I_1 = [p, 1]$ and $I_{0^k1} = \phi_0^k(I_1)$. We call ℓ_ϵ the widths of these intervals. Now the map is linearized in each interval: the corresponding slopes will be $s(I_1) = \ell_1^{-1} = p/(1-p) = \Lambda$, while $s(I_{0^k1}) = \ell_{0^{k-1}1}/\ell_{0^k1}$: the corresponding cycles stabilies will be $\Lambda_1 = \Lambda$, $\Lambda_{0^k1} = \ell_{0^k1}^{-1}$. The asymptotic behavior of ℓ_{0^k1} is determined by the intermittency exponent z, in fact if we put $y_0 = 1$, $y_1 = p$, $y_l = \phi_0(y_{l-1})$ for $l = 2, 3, \ldots$ we have that $y_{m-1} = y_m + y_m^z$ and $\delta\left(-1/y_m^{(z-1)}\right) \sim \delta m$ so that

$$y_m \sim \frac{q}{k^\alpha} \qquad \alpha = 1/(z-1)$$

This implies that the widths will scale as $\ell_{0^k1} \sim q/k^{\alpha+1}$. We now assume the former as an exact expression (and thus convert the original map to

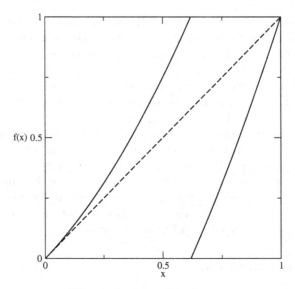

f(x) 0.5

Fig. 4. An intermittent map.

a piecewise linear approximation [22], see Fig. 5), so that $\ell_1 = \Lambda^{-1}$ and $\ell_{0^k 1} = q/k^{\alpha+1}$ and choose q in such a way that probability conservation is respected:

$$\ell_1 + \sum_{n=1}^{\infty} \ell_{0^n 1} = 1 = \Lambda^{-1} + q \cdot \zeta(1 + \alpha)$$

so $q = (\Lambda - 1)/(\Lambda \cdot \zeta(1 + \alpha))$ and the stabilities are finally written as $\Lambda_1 = \Lambda$, while $\Lambda_{0^k 1} = \Lambda \cdot \zeta(1 + \alpha) \cdot k^{\alpha+1}/(\Lambda - 1)$.

The dynamical zeta function, within the piecewise linear approximation, picks up only contributions coming from elementary cycles (corresponding to the alphabet's letters)

$$\zeta_0^{-1}(z) = 1 - \frac{z}{\Lambda} - \frac{(\Lambda - 1)}{\Lambda \cdot \zeta(\alpha + 1)} z \cdot g_{\alpha+1}(z) \tag{2}$$

where

$$g_s(z) = \sum_{m=1}^{\infty} \frac{z^m}{m^s}$$

$(g_s(z) = F(z, s)$ is usually called the Jonquière function in the mathematical litterature), while

$$\zeta_l^{-1}(z) = 1 - \frac{z}{\Lambda^l} - \left(\frac{(\Lambda - 1)}{\Lambda \cdot \zeta(\alpha + 1)} \right)^l z \cdot g_{l \cdot (\alpha+1)}(z) \tag{3}$$

From (2) we observe that, while by construction probability conservation is guaranteed ($\zeta_0^{-1}(1) = 0$), the behavior of $\zeta_0^{-1}(z)$ as $z \sim 1^-$ is not necessarily

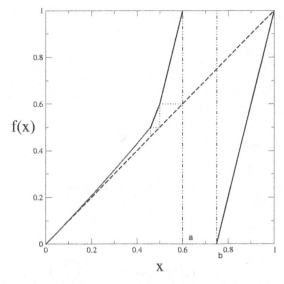

Fig. 5. A piecewise linear approximation to an intermittent map.

that of a simple zero. For instance $g_1(z) = -\log(1 - z)$, which diverges as $z \mapsto 1^-$. Now if $0 < s < 1$ again $g_s(z)$ diverges as $z \mapsto 1^-$, with a behavior like $g_s(z) \sim \Gamma(1 - s) \cdot (1 - z)^{s-1}$ (notice that however such cases are not directly relevant to the dynamical zeta function). There are a number of ways in which the former estimates can be obtained: one of the quickest procedures is to employ a Tauberian theorem for power series (we refer to [23] for proofs): let $q_n \geq 0$ and suppose that $Q(y) = \sum_{n=1}^{\infty} q_n y^n$ converges for $0 \leq y < 1$: then if $\{q_n\}$ is monotonic and

$$q_n \sim \frac{1}{\Gamma(\rho)} n^{\rho-1} \qquad n \mapsto \infty$$

with $\rho \in \mathbf{R}_+$, then

$$Q(y) \sim \frac{1}{(1 - y)^\rho} \qquad y \mapsto 1^-$$

When $s > 1$ (which is the case relevant to dynamical zeta functions (2)) the limit $(z \mapsto 1^-)$ yields a finite result $(\zeta(s))$: to see how this limit is approached we consider the representation

$$g_s(z) = \frac{z}{\Gamma(s)} \int_0^\infty d\xi \, \frac{\xi^{s-1}}{e^\xi - z}$$

which provides an analytic continuation of $F(z, s)$ over the whole complex plane, excluding the ray $(0, \infty)$, from which we can write

$$\zeta(s) - g_s(z) = \frac{(1 - z)}{\Gamma(s)} \int_0^\infty d\xi \, \frac{\xi^{s-1} e^\xi}{(e^\xi - 1) \cdot (e^\xi - z)} \tag{4}$$

Now if $s > 2$ we have

$$\int_0^\infty d\xi\, \xi^{s-1} \frac{e^\xi}{(e^\xi - 1)^2} = \Gamma(s) \cdot \zeta(s-1)$$

so that $\zeta(s) - F(z, s) \sim \zeta(s-1) \cdot (1 - z)$

If $s \in (1, 2]$ we have from (4)

$$\zeta(s) - g_s(z) = (1 - z) \sum_{m=0}^\infty z^m \zeta(s, m+1)$$

where $\zeta(s, a) = \sum_{n=0}^\infty (n + a)^{-s}$: by taking the leading term of $\zeta(s, m+1) \sim (m+1)^{s-1}$ we get

$$\zeta(s) - g_s(z) \sim (1 - z) g_{s-1}(z)$$

from which we get the dominant contribution as $z \mapsto 1^-$

$$\zeta(2) - g_2(z) \sim (1 - z) \log(1 - z) \tag{5}$$

and

$$\zeta(s) - g_s(z) \sim (1 - z)^{s-1} \tag{6}$$

for $s \in (1, 2)$. The main lesson so far is that the sequence of longer and longer cycles shadowing the marginal fixed point changes dramatically the analytic features of the dynamical zeta function: singularities develop in the form of branch points: when dealing with anomalous diffusion we will show how this leads to physically relevant consequences. Appearance of the Jonquière function makes it manifest that such dynamical zeta functions cannot be extended meromorphically outside the unit disk: z=1 is a branch point and typically a cut along the ray $(1, \infty)$ has to be imposed.

2.2 Anomalous Diffusion

Our model example is the map shown in Fig. 6, whose corresponding map on the torus is shown in Fig. 7.

Branches whose support is in I_i $i = 1, 2, 3, 4$ have uniform slope (absolute value Λ), while $f|_{I_0}$ is of intermittent form (and we will consider as a model the piecewise linear approximation formerly dealt with). Once the $\bar{0}$ fixed point is pruned away the symbolic dynamics is determined by the countable alphabet $\{1, 2, 3, 4, 0^i 1, 0^j 2, 0^k 3, 0^l 4\}$ $i, j, k, l = 1, 2, \ldots$. The partitioning of the subinterval I_0 is induced by $I_{0^k (right)} = \phi_{(right)}^k (I_3 \bigcup I_4)$ (where $\phi_{(right)}$ denotes the inverse of the right branch of $\hat{f}|_{I_0}$) and the same reasoning applies to the leftmost branch. These are regions over which the slope of $\hat{f}|_{I_0}$ is constant. Thus we have the following stabilities and weights associated to letters:

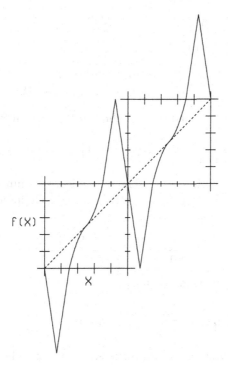

Fig. 6. An intermittent diffusing map.

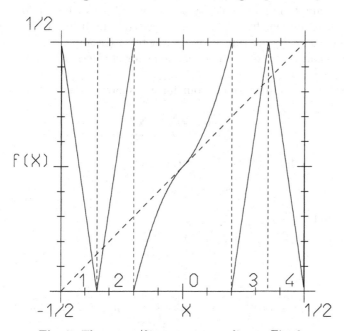

Fig. 7. The torus /f5map corresponding to Fig. 6.

$$0^k3, 0^k4 \quad \Lambda_p = \frac{k^{1+\alpha}}{q/2} \quad \sigma_p = 1$$

$$0^l1, 0^l2 \quad \Lambda_p = \frac{l^{1+\alpha}}{q/2} \quad \sigma_p = -1$$

$$3, 4 \quad \Lambda_p = \pm\Lambda \quad \sigma_p = 1$$

$$2, 1 \quad \Lambda_p = \pm\Lambda \quad \sigma_p = -1 \tag{7}$$

where q is to be determined by probability conservation for f:

$$\frac{4}{\Lambda} + 2q\zeta(\alpha+1) = 1$$

so that $q = (\Lambda - 4)/2\Lambda\zeta(\alpha + 1)$. The dynamical zeta function picks up only contributions from the alphabet's letters, as we have imposed piecewise linearity, and is written as

$$\zeta_{\beta,(0)}^{-1}(z,\beta) = 1 - \frac{4}{\Lambda}z\cosh\beta - \frac{\Lambda-4}{\Lambda\cdot\zeta(1+\alpha)}z\cosh\beta \cdot g_{\alpha+1}(z) \tag{8}$$

and its first zero $z(\beta)$ is determined by

$$\frac{4}{\Lambda}z + \frac{\Lambda-4}{\Lambda\cdot\zeta(1+\alpha)}z\cdot g_{\alpha+1}(z) = \frac{1}{\cosh\beta}$$

By using implicit function derivation we see that D vanishes (*i.e.* $z''(\beta)|_{\beta=0} = 0$) when $\alpha \leq 1$. This is easily interpreted from a physical point of view, as marginal stability implies that a typical orbit will be sticked up for long times near the $\bar{0}$ indifferent fixed point, and the 'trapping time' will be larger for higher values of the intermittency parameter z (recall $\alpha = (z-1)^{-1}$). This requires looking in more detail at the behavior of traces of high powers of the transfer operator.

In order to get deviations from linear behavior we need to recall that (6,11)

$$\left\langle e^{\beta(\hat{x}_n - \hat{x})} \right\rangle \sim \sum_{j=0}^n \lambda_j^n \tag{9}$$

Now if we introduce the *spectral determinant*

$$F_\beta(z) = \det(1 - z\mathcal{L}_\beta) \tag{10}$$

we have that

$$\frac{d}{ds}\ln F_\beta(e^{-s}) = \sum_{j=0}^\infty \frac{\lambda_j e^{-s}}{1 - \lambda_j e^{-s}} \tag{11}$$

so that

$$\mathcal{G}_n(\beta) = \left\langle e^{\beta(\hat{x}_n - \hat{x})} \right\rangle \sim \frac{1}{2\pi i}\int_{a-i\infty}^{a+i\infty} ds\, e^{sn}\frac{d}{ds}\ln F_\beta(e^{-s}) \tag{12}$$

and, in the case of one-dimensional diffusion we thus have

$$D = \lim_{t \to \infty} \frac{d^2}{d\beta^2} \left(\frac{1}{2\pi i} \int_{a-i\infty}^{a+i\infty} ds\, e^{st} \frac{F'_\beta(e^{-s})}{F_\beta(e^{-s})} \right)_{\beta=0} \tag{13}$$

As we are interested in the leading singularity we may use zeta functions instead of spectral determinants [9], thus getting

$$\langle (\hat{x}_n - \hat{x}_0)^2 \rangle \sim 2n \frac{d^2}{d\beta^2} \left(\frac{1}{2\pi i} \int_{a-i\infty}^{a+i\infty} ds\, e^{st} \frac{\partial_s \zeta^{-1}_{\beta,(0)}(e^{-s})}{\zeta^{-1}_{\beta,(0)}(e^{-s})} \right)_{\beta=0} \tag{14}$$

The evaluation of inverse Laplace transforms for high values of the argument is most conveniently performed (if applicable) by using Tauberian theorems: in particular we will employ the following version [23]: take

$$\omega(\lambda) = \int_0^\infty dx\, e^{-\lambda x} u(x)$$

(with $u(x)$ monotone in some neighborhood of infinity): then, as $\lambda \mapsto 0$ and $x \mapsto \infty$ respectively (and $\rho \in (0, \infty)$,

$$\omega(\lambda) \sim \frac{1}{\lambda^\rho} L\left(\frac{1}{\lambda}\right)$$

if and only if

$$u(x) \sim \frac{1}{\Gamma(\rho)} x^{\rho-1} L(x)$$

where L denotes any slowly varying function (*i.e.* such that $\lim_{t \to \infty} L(ty)/L(t) = 1$).

The asymptotic behavior of the integrand in (14) may then be evaluated from our estimate the behavior of Jonquière functions near $z = 1$. Omitting prefactors (which can however be calculated by the same procedure) we have

$$\frac{d^2}{d\beta^2} \frac{\partial_s \zeta^{-1}_{\beta,(0)}(e^{-s})}{\zeta^{-1}_{\beta,(0)}(e^{-s})} \Bigg|_{\beta=0} \sim \begin{cases} s^{-2} & for\ \alpha > 1 \\ s^{-(\alpha+1)} & for\ \alpha \in (0,1) \\ 1/(s^2 \log s) & for\ \alpha = 1 \end{cases}$$

from which we get the estimates (see [24, 25]

$$\langle (x_t - x_0)^2 \rangle \sim \begin{cases} t & for\ \alpha > 1 \\ t^\alpha & for\ \alpha \in (0,1) \\ t/\ln t & for\ \alpha = 1 \end{cases} \tag{15}$$

2.3 Probabilistic Approximation

In the above we used piecewise linear approximations in order to investigate dynamical implications of marginal fixed points. We now want to describe how probabilistic methods may also be employed, in order to write down approximate dynamical zeta functions for intermittent systems. The approach we are going to describe is thoroughly discussed in [26], generalizing ideas from [27].

The main point in this probabilistic approach is to suppose that any orbit might be partitioned according to a sequence of times (the idea works essentially in the same way for both discrete and continuous time systems) $t_1 < t_2 < \cdots < t_n < \cdots$ such that the time laps $\Delta_j = t_j - t_{j-1}$ form a sequence of random variables with common distribution $\psi(\Delta)\,d\Delta$ and the orbit properties before and after t_n are independent of n. A typical choice for $\{t_j\}$ in the case of intermittent maps consists in collecting the reinjection times in the laminar region: the second property mentioned above is thus related to the "randomization" operated by the chaotic phase (while such a statement is hard to control over purely theoretical bases, careful numerical checks that illustrate its validity are contained in [28]).

Before discussing how to implement this approach, we give a few examples on how an approximate form of $\psi(T)$ is determined. First of all consider the intermittent map (1): we turn it into a differential equation [29] (in I_0)

$$\dot{x}_t = x_t^z \tag{16}$$

whose solution we write as

$$x_t = \left[\frac{1}{x_0^{z-1}} - (z-1)t\right]^{-\frac{1}{z-1}} \tag{17}$$

From (17) we can obtain the exit time $T(x_0)$ for each $x_0 \in I_0$, as x_t will exit I_0 as soon as $x_t \geq p$:

$$T(x_0) = \frac{1}{z-1}\left[\frac{1}{x_0^{z-1}} - \frac{1}{p^{z-1}}\right].$$

Now

$$\psi(T) = \int_0^p dx\, P(x, T - T(x)), \tag{18}$$

where $P(x,t)$ is the probability of being injected at time t from I_1 to $x_0 \in I_0$ (we partition the time sequence so that t_j is the entrance time in the chaotic region). In the limit of large T, the dominant contribution to (18) comes from

$$\psi(T) \propto \left|\frac{dx_0(T)}{dT}\right|$$

if we suppose that the reinjection probability is smooth and chaotic residence times vanish sufficiently fast. Within this approximation we thus get

$$\psi(T) \propto \left((z-1)T + \frac{1}{p^{z-1}}\right)^{-\frac{z}{z-1}} \qquad T \gg 0$$

so that

$$\psi(T) \sim \frac{1}{T^{\frac{z}{z-1}}}$$

for large values of T.

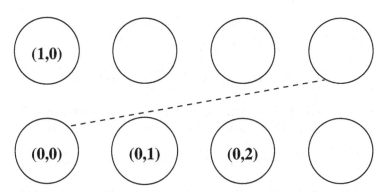

Fig. 8. Unbounded horizon Lorentz gas: the labelling is indicated for a few discs.

Now consider the periodic Lorentz gas with unbounded horizon (see Fig. 8): we will focus our attention on square lattice geometry, with disk radius R and unit lattice spacing. We label disks according to the (integer) coordinates of their center: the sequence of times $\{t_j\}$ is given by the set of collision times. We focus our attention on orbits that leave the disk sitting at the origin and hit a disk far away after a free flight. While, depending on the value of R, there may be different directions of infinite free flight, we concentrate our attention on horizontal motion. Initial conditions are characterized by coordinates (ϕ, α) (ϕ determines the initial position along the disk, while α gives the angle of the initial velocity with respect to the outward normal: the appropriate measure is then $d\phi \cos \alpha \, d\alpha$ ($\phi \in [0, 2\pi)$, $\alpha \in [-\pi/2, \pi/2]$. Our aim is to find how $\psi(T)$ scales for large values of T: this is equivalent to investigating the scaling of portions of the phase space that lead to a first collision with disk $(n, 1)$, for large values of n (as $n \mapsto \infty$ $n \simeq T$). The condition assuring that a trajectory indexed by (ϕ, α) hits the (m, n) disk (all other disks being transparent) is written as

$$\left| \frac{d_{m,n}}{R} \sin(\phi - \alpha - \theta_{m,n}) + \sin \alpha \right| \leq 1 \qquad (19)$$

where $d_{m,n} = \sqrt{m^2 + n^2}$ and $\theta_{m,n} = \arctan(n/m)$. For very small R the argument of the first term in (19) must be small and the condition simplifies to

$$\left| \frac{d_{m,n}}{R}(\phi - \alpha - \theta_{m,n}) + \sin \alpha \right| \leq 1 \qquad (20)$$

No we call j_n the portion of the phase space leading to a first collision with disk $(1, n)$: we cannot estimate j_n by simply employing (19) or (20) as part of the allowed trajectories will be screened by disks $(0, 1)$ or $(1, n-1)$ (collisions with disk $(1, n-k)$ are taken into account automatically). We thus consider $J_n = \bigcup_{k=n+1}^{\infty} j_k$: this phase space region is delimited by

$$(\phi - \alpha)_{min} = R(1 - \sin \alpha) \tag{21}$$

(tangency to upper half of $(0, 1)$ disk), and

$$(\phi - \alpha)_{max} = \theta_{n,1} - \frac{R}{d_{n,1}}(1 + \sin \alpha) \tag{22}$$

(tangency to lower half of $(1, n)$ disk). Obviously we must have that $(\phi - \alpha)_{min} \leq (\phi - \alpha)_{max}$: the (minimal) critical value of α is determined by

$$R(1 - \sin \alpha_{cr}) = \theta_{n,1} - \frac{R}{d_{n,1}}(1 + \sin \alpha_{cr})$$

As $n \mapsto \infty$ $\alpha_{cr} = \pi/2 - \delta\alpha$, where the former equation leads to the scaling $(\delta\alpha)^2 \sim 1/n$. For $\alpha = \pi/2$, from (21) and (22) we have that

$$(\phi - \pi/2)_{max} - (\phi - \pi/2)_{min} \sim 1/n$$

and thus, approximating the integration region with a triangle we get $J_n \sim 1/n^2$, and, by differentiating, $j_n \sim 1/n^3$, so that the asymptotic behavior $\psi(T) \sim T^{-3}$ is induced (a more detailed analysis of $\psi(T)$ for the Lorentz gas with infinite horizon is contained in [30].

We now return to our original question: how to use $\psi(T)$ to build a probabilistic approximation to dynamical zeta functions: let's start from the trace of the generalized evolution operator (14):

$$tr\mathcal{L}_w^t = \frac{1}{2\pi i} \int_{a-i\infty}^{a+i\infty} ds\, e^{st} \frac{\partial_s \zeta_w^{-1}(e^{-s})}{\zeta_w^{-1}(e^{-s})} = \int dx\, w(x,t)\delta(x - f^t x)$$

(where w is a generic multiplicative weight along the trajectory). If the system is ergodic (and the phase space volume is normalized to one) we can rewrite $tr\mathcal{L}_w^t$ as a time average

$$tr\mathcal{L}_w^t = \ll w(x(\tau), t)\, \delta(x(\tau) - x(t + \tau)) \gg_\tau$$

Now we write this average as a series over contributions conditioned that $\tau \in \Delta_n$ and $\tau + t \in \Delta_{n+m}$:

$$tr\mathcal{L}_w^t = \langle T \rangle \sum_{m=0}^{\infty} tr\mathcal{L}_{w:m}^t$$

$$= \langle T \rangle tr\mathcal{L}_{w:0}^t + \langle T \rangle \sum_{m=1}^{\infty} \int dz_1 \int dz_2 \int_0^{\infty} du \int_0^{\infty} dv\, z_1 p_+^w(z - 1, u) \cdot$$

$$\cdot z_2 p_-^w(z_2, v)\, (\psi W)^{*(m-1)}(t - u - v) \tag{23}$$

where $*(n)$ denotes n-fold involution, and where $p_+^w(z, u)$ is the probability that $w(x_t, u) = z$ and $u + t$ is the next exit time, $p_-^w(z, u)$ is the probability that $w(x_{t-v}, v) = z$ and the present interval has been entered since time v, while $W(t)$ is the weight associated with $\Delta = t$ (and it is assumed to depend only on the size of the time interval). In the above we notice that the factor $\langle T \rangle$ accounts for the fact that "renewal averages" are time averages sampled with $\langle T \rangle$ spacing. Now we neglect $tr\mathcal{L}_{w;0}^t$ (which takes care of events happening within the same interval Δ_i) and take the Laplace transform of the right hand side of (23): we obtain the identity

$$\frac{\partial_s \zeta_w^{-1}(e^{-s})}{\zeta_w^{-1}(e^{-s})} = \frac{\int_0^\infty dt\, e^{-st} \int_0^t du \int dz_1 \int dz_2\, z_1 p_+^w(z_1, u) z_2 p_-^w(z_2, t - u)}{1 - \int_0^\infty dt\, e^{-st} W(t)\psi(t)} \quad (24)$$

where we have used the convolution theorem for Laplace transforms. Now, since

$$\int_0^t du \int dz_1 \int dz_2\, z_1 p_+^w(z_1, u) z_2 p_-^w(z_2, t - u)$$

may be identified with $t\psi(t)W(t)$, we can consistently put

$$\zeta_w^{-1(prob)}(e^{-s}) = 1 - \int_0^\infty dt\, e^{-st} W(t)\psi(t) \quad (25)$$

which gives the probabilistic approximation to the dynamical zeta function. Computations using this probabilistic zeta function for the Lorentz gas with infinite horizon are provided in [30].

2.4 Resident Times and Correlation

In dealing with generic dynamical systems one cannot easily recognize the "randomizing mechanism", thus no evident way of building a $\psi(T)$ is a priori given. Nevertheless use of time laps distributions have proven to be quite useful since a long time [31–33]. While the method may be formulated essentially in the same form for continuous and discrete time dynamics, we illustrate it for the discrete case: the extention of our formulas to the continuous time case is straightforward. We consider the dynamical evolution S acting on the phase space $\mathcal{F}$ which we divide into two parts, $\mathcal{F}_0$ and $\mathcal{F}_1$, and then run a long trajectory $x_0, x_1, x_2, \ldots$ from which we extract a sequence of residence times in the same subregion $t_0, t_1, \ldots$, where $t_j = m$ means that in the original sequence there is a substring $x_j, x_{j+1}, \ldots x_{j+m-1}$ in the same $\mathcal{F}_\epsilon$ while x_{j-1} and x_{j+m} belong to the complementary subregion. From the t sequence we may reconstruct an approximate residence times probability distribution $\mu(n)$, which we suppose gives a finite average residence time $\langle l \rangle = \sum m \cdot \mu(m)$. Then, if we refer to the original sequence $x_0, x_1, x_2, \ldots$, we may evaluate the probability that a point chosen at random is the starting point of an m–long residence sequence as $\mu(m)/\langle l \rangle$ (this is also equal to the probability that the point is the *last* point of an m-long residence sequence), while the probability that a point chosen at random simply belongs to an m-long sequence is $m \cdot \mu(m)/\langle l \rangle$. We can also define integrated probabilities as follows:

$$P_{int}(m) = \sum_{k \geq m} \mu(k) \tag{26}$$

that is the probability that a residence sequence is at least long m.

The intuitive argument that suggests a relationship between $P_{int}(m)$ and correlation decay is as follows: we introduce a function $\tilde{C}_m$ defined as the probability that two points, m iterations apart, chosen at random, belong to the same residence subsequence. While this quantity is an indicator of memory effects in dynamical evolution it is not formally defined as a correlation function of an observable, and thus our arguments are intuitive rather than rigorous. We now express $\tilde{C}_m$ in terms of P_{int}:

$$\tilde{C}_m = \frac{1}{\langle l \rangle} \left(\mu(m) + 2 \cdot \mu(m+1) \ldots \right) \tag{27}$$

and thus, by performing partial summations,

$$\tilde{C}_m = \frac{1}{\langle l \rangle} \sum_{k \geq m} P_{int}(k) \tag{28}$$

Quantitatively this means that if $P_{int}(m)$ decays exponentially, the same holds for $\tilde{C}_m$, while a power-law decay

$$P_{int}(m) \sim m^{-(\alpha+1)} \tag{29}$$

induces the behaviour $\tilde{C}_m \sim m^{-\alpha}$.

While the former considerations are based on the *interpretation* of C_m as a correlation function we remark that the validity of the technique is strongly confirmed by rigorous arguments (that cover exponential as well as power-law mixing [34]: though these papers are concerned with *return times* statistics, we believe that the kind of approach may be generalized to other definitions of time intervals). Indeed the technique may be tested in numerical experiments where theoretical results are *a priori* known: for instance if one considers the motion of a test particle in a Lorentz gas with infinite horizon (see [35, 36]), then strong theoretical arguments predict a $1/t$ decay asymptotic decay of correlations: the integrated probability of disk to disk collisions shows accordingly a $1/t^2$ decay law (see Fig. 9). Further checks of the same nature may be performed also for other classes of billiard tables [37].

This work was partially supported by the PRIN-2000 project *Chaos and Localization in classical and quantum systems*, and by INFM, PA 2002 *Weak chaos, theory and applications*.

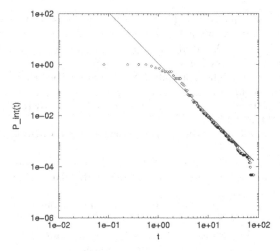

Fig. 9. Collision time statistics for the infinite horizon Lorentz gas. The straight line has slope -2.

References

1. Klages, R., Dorfman, J.R. (1995): Simple maps with fractal diffusion coefficient. Phys. Rev. Lett., **74**, 387–390
2. Artuso R. (1991): Diffusive dynamics and periodic orbits of dynamical systems. Phys. Lett., **A 160**, 528–530
3. Cvitanović, P., Gaspard, P., Schreiber T. (1992): Investigation of the Lorentz Gas in terms of periodic orbits. CHAOS, **2**, 85–90
4. Cvitanović, P., Eckmann, J.-P., Gaspard P. (1995): Transport properties of the Lorentz gas in terms of periodic orbits. Chaos, Solitons and Fractals, **6**, 113–120
5. Dana, I. (1989): Hamiltonian transport on unstable periodic orbits. Physica, **D 39**, 205–230
6. Vance, W.N. (1992): Unstable periodic orbits and transport properties of nonequilibrium steady states. Phys. Rev. Lett., **96**, 1356–1359
7. Artuso, R., Aurell, E., Cvitanović, P. (1990): Recycling of strange sets I: Cycle expansions. Nonlinearity **3**, 325–360
8. Artuso, R., Aurell, E., Cvitanović, P. (1990): Recycling of strange sets II: Applications. Nonlinearity **3**, 361–386
9. Cvitanović, P., Artuso, R., Mainieri, R., Tanner, G., Vattay, G. (2001): Classical and Quantum Chaos. www.nbi.dk/ChaosBook/, Niels Bohr Institute, Copenhagen
10. Baladi, V. (1995): Dynamical zeta functions. In: Branner, B., Hjorth (eds) Proceedings of the NATO ASI Real and Complex Dynamical Systems. Kluwer Academic Publishers Dordrecht
11. Devaney, R.L. (1987): An Introduction to Chaotic Dynamical Systems. Addison-Wesley Reading MA
12. Hansen, K.T. (1994): Symbolic Dynamics in Chaotic Systems. Ph.D. thesis, University of Oslo www.nbi.dk/CATS/papers/khansen/thesis/thesis.html

13. Artuso, R., Strepparava, R. (1997): Recycling diffusion in sawtooth and cat maps. Phys. Lett. **A236** 469–475
14. Arnol'd V.I., Avez, A. (1967): Problèmes Ergodiques de la Mécanique Classique. Gauthier-Villars Paris
15. Rugh, H.H. (1992): The Correlation Spectrum for Hyperbolic Analytic Maps. Nonlinearity **5**, 1237–1263
16. Percival, I., Vivaldi, F. (1987): A linear code for the sawtooth and cat maps. Physica **27D**, 373–386
17. Bird, N., Vivaldi, F. (1988): Periodic orbits of the sawtooth maps. Physica **D30**, 164–176
18. Coxeter, H.S.M. (1948): Regular Polytopes. Methuen London
19. Lichtenberg, A.J., Lieberman, M.A. (1982): Regular and stochastic motion. Springer New York
20. Cary, J.R., Meiss, J.D. (1981): Rigorously diffusive deterministic map. Phys. Rev. **A24**, 2664–2668
21. Pomeau, Y., Manneville, P.(1980): Intermittent transition to turbulence in dissipative dynamical systems. Commun. Math. Phys. **74**, 189–197
22. Gaspard, P., Wang, X.-J. (1988): Sporadicity: between periodic and chaotic dynamical behavior. Proc. Natl. Acad. Sci. U.S.A. **85** 4591–4595
23. Feller, W. (1966): An introduction to probability theory and applications, Vol. II. Wiley New York
24. Geisel, T., Thomae, S. (1984): Anomalous diffusion in intermittent chaotic systems. Phys. Rev. Lett. **52** 1936–1939
25. Artuso, R., Casati G., Lombardi, R. (1993): Periodic orbit theory of anomalous diffusion. Phys. Rev. Lett. **71** 62–64
26. Dahlqvist, P. (1995): Approximate zeta functions for the Sinai billiard and related systems. Nonlinearity **8**, 11–28
27. Baladi, V., Eckmann, J.-P., Ruelle, D. (1989): Resonances for intermittent systems. Nonlinearity **2**, 119–135
28. Dahlqvist, P., Artuso, R. (1996): On the decay of correlations in the Sinai billiard with infinite horizon. Phys. Lett. **A 219**, 212–216
29. Procaccia, I., Schuster, H. (1983): Functional renormalization group theory of universal 1/f noise in dynamical systems. Phys. Rev. **A 28**, 1210–1212
30. Dahlqvist, P. (1996): Lyapunov exponents and anomalous diffusion of a Lorentz gas with infinite horizon. J. Stat. Phys. **84** 773–795
31. Karney, C.F.F. (1983): Long-time correlations in the stochastic regime. Physica **D 8**, 360–380
32. Chirikov, B.V., Shepelyansky, D.L. (1984): Correlation properties of dynamical chaos in Hamiltonian systems. Physica **D 13**, 395–400
33. Artuso, R. (1999): Correlation decay and return time statistics. Physica **131** 68–77
34. Young, L.-S. (1999): Recurrence times and rates of mixing. Israel J. Math. **110** 153–188
35. Bunimovich, L.A. (1985): Decay of correlations in dynamical systems with chaotic behavior. Sov. Phys. JETP **62** 842–852
36. Bleher, P.M. (1992): Statistical properties of two-dimensional periodic Lorentz gas with infinite horizon. J. Stat. Phys. **66** 315–373
37. Artuso, R., Casati, G., Guarneri, I. (1996): Numerical experiments on billiards. J. Stat. Phys. **83** 145–166

Lecture Notes in Physics

For information about Vols. 1–577
please contact your bookseller or Springer-Verlag
LNP Online archive: http://www.springerlink.com/series/lnp/

Vol. 578: D. Blaschke, N. K. Glendenning, A. Sedrakian (Eds.), Physics of Neutron Star Interiors.

Vol. 579: R. Haug, H. Schoeller (Eds.), Interacting Electrons in Nanostructures.

Vol. 580: K. Baberschke, M. Donath, W. Nolting (Eds.), Band-Ferromagnetism: Ground-State and Finite-Temperature Phenomena.

Vol.581: J. M. Arias, M. Lozano (Eds.), An Advanced Course in Modern Nuclear Physics.

Vol.582: N. J. Balmforth, A. Provenzale (Eds.), Geomorphological Fluid Mechanics.

Vol.583: W. Plessas, L. Mathelitsch (Eds.), Lectures on Quark Matter.

Vol.584: W. Köhler, S. Wiegand (Eds.), Thermal Nonequilibrium Phenomena in Fluid Mixtures.

Vol.585: M. Lässig, A. Valleriani (Eds.), Biological Evolution and Statistical Physics.

Vol.586: Y. Auregan, A. Maurel, V. Pagneux, J.-F. Pinton (Eds.), Sound–Flow Interactions.

Vol.587: D. Heiss (Ed.), Fundamentals of Quantum Information. Quantum Computation, Communication, Decoherence and All That.

Vol.588: Y. Watanabe, S. Heun, G. Salviati, N. Yamamoto (Eds.), Nanoscale Spectroscopy and Its Applications to Semiconductor Research.

Vol.589: A. W. Guthmann, M. Georganopoulos, A. Marcowith, K. Manolakou (Eds.), Relativistic Flows in Astrophysics.

Vol.590: D. Benest, C. Froeschlé (Eds.), Singularities in Gravitational Systems. Applications to Chaotic Transport in the Solar System.

Vol.591: M. Beyer (Ed.), CP Violation in Particle, Nuclear and Astrophysics.

Vol.592: S. Cotsakis, L. Papantonopoulos (Eds.), Cosmological Crossroads. An Advanced Course in Mathematical, Physical and String Cosmology.

Vol.593: D. Shi, B. Aktaş, L. Pust, F. Mikhailov (Eds.), Nanostructured Magnetic Materials and Their Applications.

Vol.594: S. Odenbach (Ed.),Ferrofluids. Magnetical Controllable Fluids and Their Applications.

Vol.595: C. Berthier, L. P. Lévy, G. Martinez (Eds.), High Magnetic Fields. Applications in Condensed Matter Physics and Spectroscopy.

Vol.596: F. Scheck, H. Upmeier, W. Werner (Eds.), Noncommutative Geometry and the Standard Model of Elememtary Particle Physics.

Vol.597: P. Garbaczewski, R. Olkiewicz (Eds.), Dynamics of Dissipation.

Vol.598: K. Weiler (Ed.), Supernovae and Gamma-Ray Bursters.

Vol.599: J.P. Rozelot (Ed.), The Sun's Surface and Subsurface. Investigating Shape and Irradiance.

Vol.601: F. Mezei, C. Pappas, T. Gutberlet (Eds.), Neutron Spin Echo Spectroscopy. Basics, Trends and Applications.

Vol.602: T. Dauxois, S. Ruffo, E. Arimondo (Eds.), Dynamics and Thermodynamics of Systems with Long Range Interactions.

Vol.603: C. Noce, A. Vecchione, M. Cuoco, A. Romano (Eds.), Ruthenate and Rutheno-Cuprate Materials. Superconductivity, Magnetism and Quantum Phase.

Vol.604: J. Frauendiener, H. Friedrich (Eds.), The Conformal Structure of Space-Time: Geometry, Analysis, Numerics.

Vol.605: G. Ciccotti, M. Mareschal, P. Nielaba (Eds.), Bridging Time Scales: Molecular Simulations for the Next Decade.

Vol.606: J.-U. Sommer, G. Reiter (Eds.), Polymer Crystallization. Obervations, Concepts and Interpretations.

Vol.607: R. Guzzi (Ed.), Exploring the Atmosphere by Remote Sensing Techniques.

Vol.608: F. Courbin, D. Minniti (Eds.), Gravitational Lensing:An Astrophysical Tool.

Vol.609: T. Henning (Ed.), Astromineralogy.

Vol.610: M. Ristig, K. Gernoth (Eds.), Particle Scattering, X-Ray Diffraction, and Microstructure of Solids and Liquids.

Vol.611: A. Buchleitner, K. Hornberger (Eds.), Coherent Evolution in Noisy Environments.

Vol.612 L. Klein, (Ed.), Energy Conversion and Particle Acceleration in the Solar Corona.

Vol.613 K. Porsezian, V.C. Kuriakose, (Eds.), Optical Solitons. Theoretical and Experimental Challenges.

Vol.614 E. Falgarone, T. Passot (Eds.), Turbulence and Magnetic Fields in Astrophysics.

Vol.615 J. Büchner, C.T. Dum, M. Scholer (Eds.), Space Plasma Simulation.

Vol.616 J. Trampetic, J. Wess (Eds.), Particle Physics in the New Millenium.

Vol.617 L. Fernández-Jambrina, L. M. González-Romero (Eds.), Current Trends in Relativistic Astrophysics, Theoretical, Numerical, Observational

Vol.618 M. Degli Esposti, S. Graffi (Eds.), The Mathematical Aspects of Quantum Maps

Vol.619 H.M. Antia, A. Bhatnagar, P. Ulmschneider (Eds.), Lectures on Solar Physics

Vol.620 C. Fiolhais, F. Nogueira, M. Marques (Eds.), A Primer in Density Functional Theory

Vol.621 G. Rangarajan, M. Ding (Eds.), Processes with Long-Range Correlations